paper
alder
coffee

Springer Series in Behavior Modification
THEORY / RESEARCH / APPLICATION

Cyril M. Franks / Series Editor

Volume 1
Multimodal Behavior Therapy
Arnold A. Lazarus

Volume 2
Behavior-Therapy Assessment
Eric J. Mash and Leif G. Terdal, editors

Volume 3
Behavioral Approaches to Weight Control
Edward E. Abramson
with contributors

Volume 4
A Practical Guide to Behavioral Assessment
Francis J. Keefe, Steven A. Kopel, and Steven B. Gordon

Francis J. Keefe is Director of the Psychophysiology Laboratory in the Department of Psychiatry, Harvard Medical School, and Director of Clinical Training at Learning Therapies, Inc., Newton, Massachusetts. He received his B.A. from Bowdoin College, Maine, and his M.S. and Ph.D. degrees in clinical psychology from Ohio University. Dr. Keefe specializes in psychophysiology and has written a number of papers on behavior therapy and the control of physiological function and disease.

Steven Alan Kopel is Assistant Professor, in the Departments of Community Medicine and Psychiatry at Rutgers Medical School. He is also Co-Principal Investigator of Rutgers' Multiple Risk Factor Intervention Trial for the prevention of coronary heart disease. He received his B.A. from the State University of New York at Stony Brook, and his M.A. and Ph.D. degrees from the University of Oregon. Dr. Kopel is the author of many articles in the fields of behavior and sex therapy.

Steven B. Gordon, is Clinical Associate Professor of Psychiatry at Rutgers Medical School and Consulting Director of the Child and Family Center of the Institute for Behavior Therapy in New York City. Dr. Gordon received his Ph.D. degree in clinical psychology from West Virginia University. He is the author of several articles on the topic of training parents and teachers in the use of behavior therapy.

A Practical Guide to Behavioral Assessment

Francis J. Keefe
Harvard Medical School

Steven A. Kopel
Rutgers Medical School

Steven B. Gordon
Rutgers Medical School

RC469
K43
1978

 Springer Publishing Company
New York

To Delia, Roni, and Rita

Copyright © 1978 by Springer Publishing Company, Inc.

All rights reserved

No part of this publication may be reproduced, stored in a retrieval system, or transmitted, in any form or by any means, electronic, mechanical, photocopying, recording, or otherwise, without the prior permission of Springer Publishing Company, Inc.

Springer Publishing Company, Inc.
200 Park Avenue South
New York, N.Y. 10003

78 79 80 81 82 / 10 9 8 7 6 5 4 3 2 1

Library of Congress Cataloging in Publication Data

Keefe, Francis J
   A practical guide to behavioral assessment.

   (Springer series in behavior modification; v. 4)
   Bibliography: p.
   Includes index.
   1. Mental illness—Diagnosis. 2. Personality assessment. I. Kopel, Steven A., joint author. II. Gordon, Steven B., joint author. III. Title. [DNLM: 1. Personality assessment. 2. Social behavior disorders—Diagnosis. WM145 K254p]
RC469.K43   616.8'9'075   77-27067
ISBN 0-8261-2100-4
ISBN 0-8261-2101-2 pbk.

Printed in the United States of America

Contents

Foreword vii

Preface xi

PART I: PRINCIPLES

Chapter 1
A Comparison of Traditional Assessment
and Behavioral Assessment 3

Chapter 2
A Procedural Framework for Behavioral
Assessment 16

PART II: APPLICATIONS

Chapter 3
Behavioral Assessment with Children 37

Chapter 4
Behavioral Assessment with Adult Outpatients 64

Chapter 5
Behavioral Assessment of Marital Discord 91

Chapter 6
Behavioral Assessment of Sexual Dysfunction 107

Chapter 7
Behavioral Assessment
of the Institutionalized Patient   126

PART III: CURRENT PERSPECTIVES
AND FUTURE DIRECTIONS

Chapter 8
Behavioral Assessment of Systems and Society   147

Chapter 9
Future Directions   165

APPENDIXES

Appendix A
Family Questionnaire   173

Appendix B
Multimodal Life History Questionnaire   176

References   185

Index   209

Foreword

The status of psychodiagnostic testing has changed surprisingly little in the past three decades. Regrettably, it remains the poor relative of psychotherapy. To understand fully how this prestige gap came about, it would be necessary to take a hard look at two closely interwoven strands: the historical rivalry between the professions of psychiatry and clinical psychology on the one hand, and the very nature of psychodynamically-oriented psychotherapy on the other. In essence, psychotherapy was—and still is in certain psychiatric circles—regarded as the treatment of a disease process and therefore properly relegated to the domain of the fully qualified medical practitioner. Psychological testing, as with any other physician's aid, is then regarded as but one adjunctive test among many, to be "ordered" by the physician as seen fit. The link between testing and treatment is thus a weak one.

Other than at the grossest of levels, neither diagnostic classification *per se* nor test findings typically lead in themselves to differential treatment. Despite the documented unreliability of diagnosis and its questionable nosological foundations, psychodiagnostic labels often tend to carry an authority and a stigmatizing connotation that are difficult to correct. Furthermore, the traditional stress within the psychodynamic camp upon enduring personality traits and intrapsychic configurations rather than external determinants could lead to the underestimation of environmental influences as maintaining variables. It might also be noted that, since psychotherapy and diagnosis are relatively independent, it is possible to engage in one of these activities primarily, to the exclusion of the other. And certainly, for most of us, helping a person change for the better is more gratifying than doing a psychological test. Under these circumstances, and with certain significant exceptions, it is not surprising that psychotherapy received, and continues to receive, most of the prestige and glamour, leaving few rewards for its poor and rather distant relative.

Obviously, it is not possible to engage in extended discussion of these complex matters here and the present overview is, of necessity, an oversimplification. Psychological testing is but one aspect of the diagnostic process as far as psychotherapy is concerned, and not all testing is predicated upon psychodynamic concepts. And whereas testing is usually a specialty carried out only by certain clinical psychologists or similarly trained individuals, behavioral assessment is a broad-based concept that is integral to training and practice in behavior therapy regardless of professional identity.

With the above caveat in mind, and without wishing to encourage back-patting or double standards, the contrast between psychological testing and behavioral assessment would seem to be striking. Most good therapists, regardless of persuasion, recognize the need to engage in some kind of periodic stock-taking with respect to progress and procedures. But, for the behavior therapist, this is not an optional afterthought; it is mandatory. Rigorous and continual evaluation is part and parcel of the behavioral model. Systematic assessment and intervention are closely integrated, leading to a conceptual and methodological unity (not to be equated with rigidity or a closed system) that makes it virtually impossible to engage meaningfully in one without consideration of the other. Like the Roman god, Janus, these are but two faces of one entity. Potentially misleading diagnostic labels are replaced by data-oriented profiles of the individual's functional repertoire and their maintaining stimuli. The static process of psychological testing, even if carried out before and after therapy, is replaced by the ongoing dynamics of behavioral evaluation. The model becomes that of the behavioral scientist, with allegiance to what Yates calls "a methodological prescription" rather than to a set of doctrines or a professional affiliation. While some form of social learning theory is the preferred conceptual modality at the present time, the competent behavior therapist is prepared and equipped to incorporate into his or her framework knowledge derived from such related disciplines as physiology, pharmacology, systems theory, and environmental ecology as the situation demands.

No informed individual nowadays would define behavior therapy in terms of techniques, or a behavior therapist as one who has demonstrably mastered a sufficient number of these procedures. It is equally, if perhaps less obviously, erroneous to define behavioral assessment in terms of currently available procedures or to restrict the choice of such procedures to those stemming directly from S-R learning theory. Behavioral assessment, as readers of this volume will readily find out, makes use of a variety of strategies culled from a variety of sources. The common link is that it does so behaviorally. All clinicians assess behavior for some purpose or another, even if—in the case of the psychoanalyst—only to make inferences about putative underlying constructs. Few clinicians, other than behavior therapists, do so *behaviorally*, a process that entails the systematic, objective scrutiny of all pertinent responses and maintaining variables in terms of the total environmental situation.

Unfortunately, just as there are often gaps between the written description of what the behavior therapist is supposed to be doing and what he or she actually does, so it is with behavioral assessment. "Talking the talk" and "walking the walk" are not always synonymous. As interest in behavior therapy proliferates, so the demand for behavioral assessment grows. But when demand begins to exceed supply, it becomes tempting, despite our training, to accept procedures whose conceptual and methodological underpinnings—not to mention such seminal matters as validity, reliability, and the availability of normative data—are wanting. The popularity of glib journal articles and brief workshops devoted to techniques of behavioral assessment bears far from mute testimony to the viability of these observations.

Fortunately, several thoughtful and cautiously encouraging edited texts on behavioral assessment are also in evidence and it is among these that the present volume by Doctors Keefe, Kopel, and Gordon takes its well-deserved place. *A Practical Guide to Behavioral Assessment* has the distinction of being among the first—if not the first—major, newly written texts on this topic to appear. The authors' frame of reference for the assessment of behavior is the entire compass of behavior therapy. Circumspection and scientific rigor are balanced by a forward-looking openness to innovation. Principles, applications, current perspectives, and future directions are skillfully blended. Most important, these wise clinicians stress throughout the total sociopsychological context within which their patients live. Not for them the single-minded, single-modality framework for which early behavior therapists were so rightly taken to task!

The *Practical Guide* is by no means a facile or uncritical tract. Limitations as well as strengths are underscored. Those in search of instant recipes, bereft of any need to appreciate what behavior therapy is all about, will probably be disappointed. While major assessment techniques are indeed enumerated in step-by-step detail, this is no cookbook. For discerning, discriminating readers seeking to incorporate the best of behavioral assessment into their behavior therapy armamentarium, this book is likely to be invaluable. It is indeed, as the authors promise, a practical guide to behavior therapy. But it is more—it represents a wave of the future, a working model for others to follow in the years to come.

CYRIL M. FRANKS
Professor of Psychology
Graduate School of Applied
  and Professional Psychology
Rutgers University

Preface

Throughout history, humankind has continued to ask the seemingly simple question "Why?" Our existence has depended on our ability to understand and cope with an often hostile world around us. Advances in science have made it possible for us to understand the environment more fully and even to formulate laws of nature. Associated technological advances have increased our ability to control the forces of nature.

The human race has never been satisfied with restricting the search for knowledge to the world around us, however. Our thoughts and actions offer an intriguing puzzle. "Crazy," "odd," or "unusual" behaviors are perplexing. Who or what causes these behaviors? Can they be changed? Do individuals engaging in these behaviors pose a threat to society?

We have always felt an urgent need to answer these questions. Early in history, religious, mystical, and magical explanations met this need. It is only recently that we have begun systematically to apply the principles of scientific inquiry to the assessment and modification of behavior. This approach has served as the basis for the field of behavior therapy.

Given the widespread attention that many procedures within behavior therapy have received, it is rather disappointing to note the relative lack of concern with behavioral assessment until very recently. Several publications have presented a framework or discussed the concept of behavioral assessment. A number of recent books on behavior therapy have either presented one chapter or small sections of chapters devoted to behavioral assessment. However, a comprehensive source aimed at the practical applications of behavioral assessment is sorely missing, and the main purpose of *A Practical Guide to Behavioral Assessment* is to fill that gap.

In this book we present a working model for behavioral assessment that is based on our own clinical experience. The book is divided into three parts. Part I focuses on the basic theoretical and practical foundations of behavioral assessment. Chapter 1 outlines the characteristics of behavioral assessment and contrasts them with traditional psychological assessment. Chapter 2 presents a procedural framework that serves to guide the clinician through the entire process of behavioral assessment from initial intake to long-term follow-up. Part II focuses on practical applications of the principles of behavioral assessment to various clinical populations. Practical issues in the assessment of children (Chapter 3), adult outpatients (Chapter 4), marital discord (Chapter 5), sexual dysfunction (Chapter 6), and institutionalized patients (Chapter 7) are discussed. In Part III we examine some of the new frontiers for behavioral assessment. Chapter 7 reviews recent developments in the assessment of society and systems, and Chapter 8 presents a glimpse of future directions for the development of behavioral assessment.

We would like to thank several of our teachers, colleagues, and friends whose wisdom, support, and encouragement have contributed to this book. We are particularly grateful to Dr. Cyril M. Franks, whose encouragement was an important initial impetus. We would like to thank Drs. Lee Birk, Marvin Goldfried, Gerald C. Davison, Nick Wiltz, Joseph LoPiccolo, and Robert L. Weiss for their contributions to our thinking about behavior therapy and assessment. We would also like to acknowledge several colleagues whose critical comments on earlier versions of the manuscript were helpful. We thank Richard Surwit, Larry Young, Joanne Hager, and Myron Gessner. Finally, we would like to extend our deepest thanks to our wives, Delia, Roni, and Rita, for their patience and support during the preparation of this book.

February 1978

Francis J. Keefe
Steven A. Kopel
Steven B. Gordon

PART I
Principles

CHAPTER 1

A Comparison of Traditional Assessment and Behavioral Assessment

Assessment is a feature of most approaches to psychotherapy. There are, however, certain commonalities among traditional approaches to assessment that distinguish them from the behavioral approach. Section one of this chapter compares and contrasts the theory, uses, and methods of behavioral assessment and traditional assessment; section two examines their relative strengths and weaknesses; and section three considers the implications of adherence to the behavioral approach for the practicing clinician.

Theoretical Foundations, Uses, and Methods of Traditional Assessment and Behavioral Methods

This section compares traditional assessment and behavioral assessment regarding: (1) theoretical foundations; (2) uses of assessment; and (3) methods of assessment.

Theoretical Foundations of Traditional Assessment

Psychodynamic theory and trait theory form the theoretical base for traditional approaches to assessment. Both view personality as the central factor in understanding, predicting, or changing behavior. Psychodynamic theory views emotional disturbance as the product of underlying conflicts between inferred forces (psychodynamics) such as needs, drives, or motives. Trait theory maintains that an individual acts a certain way because he has a particular trait or personality characteristic such as aggressiveness, dependency, or honesty. Both psychodynamic theory and trait theory view the pychodynamics or traits as consistent, stable, general causes of behavior. Thus, they posit that inferred dispositions determine behavior in a consistent, pervasive manner over time and across different situations or settings (e.g., Cattell, 1950; Sanford, 1963).

Both traditional theories also take a *sign* approach to behavior. Observable behaviors provide signs for inferred global personality characteristics. In the trait model, behaviors are viewed as *direct* signs of the personality trait, with the strength of the trait determined by the number of signs—an additive model. Importantly, the trait label is considered not merely a descriptor, but rather is seen as an underlying personality characteristic that *causes* the observable behaviors (signs) (Allport, 1966). In the psychodynamic model, behaviors are *indirect* signs (or symbols) of an underlying personality organization and dynamic forces; any aspect of behavior, for example, fears or obsessive thoughts, may be a symbol that reveals the underlying dynamics.

Trait theory and psychodynamic theory depend upon many levels of inference. This is apparent in traditional psychological testing procedures (Goldfried & Kent, 1972). In general, the major inferences made in testing are that the test performance: (1) is based upon representative samples of test stimuli; (2) is a valid sign of either traits or underlying psychodynamics; and (3) reveals general and stable causes of thoughts, feelings, and behavior outside the test situation.

Traditional therapists maintain that a thorough understanding of underlying dynamics or traits is needed to design effective therapy. Treatment focuses on changing the underlying causes of psychological problems through insight.

Theoretical Foundations of Behavioral Assessment

Social learning theory (e.g., Bandura, 1969; Kanfer & Phillips, 1970) provides the theoretical foundation of behavioral assessment. The focus is on *behavior*—how the person acts, what he does in a variety of life situations. Behaviors are seen as being acquired and maintained according to established principles of learning, regardless of whether they are labeled "normal" or "abnormal" by society. Social learning theory stresses the interrelationships between behavior and the environment.

The theoretical model of social learning theory has undergone many revisions over the years. In the 1920s, radical behaviorists such as John B. Watson maintained that behavior could be understood using a simple S-R model:

S———————————————————R
Stimuli   *Responses*
environmental events   muscular movements
or physiological states   or glandular secretions

This reductionistic model is viewed as a "black-box" model because of its mechanistic approach. The model attends only to inputs (stimuli) and outputs (responses). It assumes automatic links between these two elements. "Mental" events are denied independent status and reduced to physiochemical processes.

Research in learning during the 1930s and 1940s raised serious questions about the adequacy of the S-R model. Other variables in addition to stimuli and responses were seen as important to the learning process. Hull stressed intervening variables such as habit strength and drive. The role of the organism itself was recognized and a new class of variables, organismic variables, was emphasized. The work of Skinner and others contributed to the model by stressing the influence of consequences and schedules of reinforcement.

Kanfer and Saslow (1965) incorporated the various classes of variables found to influence behavior to form the S-O-R-K-C model:

S———O———R———K———C
Stimulus Organismic   Contingency
Events   Variables   Responses Relationships Consequences

Stimulus events might include: (1) physical stimuli, such as an elevator, a crowded room, or a cemetery; (2) social stimuli, such as verbal praise or attention; or (3) internal stimuli, such as fearful thoughts or heart palpitations.

Organismic variables include the biological condition of the individual, such as unique physical handicaps or drug-induced states.

Responses are viewed as occurring in one or more of the following response systems: (1) motor; (2) cognitive; and (3) physiological. These are the behaviors of the individual that occur overtly, as motoric or verbal responses, or covertly, as thinking or internal physiological reactions.

Contingency relationships describe the arrangement that exists between behavior and its consequences. The likelihood that a hypochondriac may receive social reinforcement for chronic verbal complaints is quite variable. The contingency relationship specifies the likelihood that such a response will be reinforced. This relationship has also been termed the "schedule of reinforcement." Schedules describe the frequency of consequences relative to the target response. Reinforcement may be continuous, occurring after each response. Reinforcement may be on an intermittent schedule defined by the number of responses necessary before reinforcement (known as a ratio schedule) or by the amount of time before reinforcement occurs (an interval schedule). Both ratio and interval schedules may remain constant (fixed schedule) or may vary (variable schedule).

Consequences are those events, both positive and negative, that follow the responses. Consequences may be physical events, social events, or even self-generated events, such as rewarding oneself after accomplishing a goal.

The key to behavioral assessment is an analysis of the specific variables that are controlling or affecting behavior in specific situations. This approach, to discover the relationships between behavior and environment, has been called a *behavioral-analytic strategy*. It leads to a *functional analysis* in terms of the S-O-R-K-C elements of the model (Ferster, 1965; Kanfer & Saslow, 1965; Goldfried & D'Zurilla, 1969). Thus, the strategy of behavioral assessment is based upon an empirical model that sets out to discover functional relationships through scientifically sound procedures. The search for controlling or maintaining variables is guided by two major principles: the principle of direct sampling and the principle of operational definitions.

The principle of direct sampling. Problematic behaviors are not viewed as signs of underlying disturbance, but rather, as the problem itself. Behavioral assessment requires an adequate sampling of environmental events, behaviors, and consequences in the settings where the behavior is problematic or maladaptive. The descriptions of the conditions or situations under which the behavior occurs, as well as the behavior itself and the consequences, are of fundamental importance.

The principle of operational definitions. The second principle is to operationalize terms. Vague or general terms are translated into operational definitions. Thus, a complaint such as "feeling depressed" might be operationally defined by the number of hours spent sleeping each day, the number of crying episodes occurring each week, or the number of different activities engaged in each month. Vague terms are redefined by the operations or events that define the term for the particular individual in question; assumptions are not based upon a descriptor label, such as "depressed." An important characteristic of an operational definition is that the terms used are measurable and observable (directly detectable) by the patient and/or others.

In summary, behavioral assessment relies upon social learning theory. Assessment is guided by the use of specific detectable units and the principle of operational definitions. Behavioral assessment aims to discover functional relationships between behavior and environment through direct sampling of relevant behaviors and the environmental conditions.

Uses of Traditional Assessment

Traditional assessment and behavioral assessment often are used for purposes not directly related to therapy treatment plans, including: (1) selection and placement; (2) administrative or statistical accounting; (3) legal decisions; and (4) research endeavors. Our major focus in this book is on assessment for the purpose of treatment.

Although there is some evidence to the contrary (e.g., Daily, 1953; Meehl, 1960), a major purpose of traditional assessment is to derive a "personality profile" in terms of psychodynamic or trait concepts that is then used to guide the clinician in the treatment phase. Traditional assessment methods may yield case formulations stressing historical factors in terms of Oedipal complexes, ego strength, or trait dispositions such as dependency or hostility.

The practice of diagnostic classification is common in traditional assessment. In many clinical settings, the case or personality formulation is summarized, or replaced, by a simple diagnostic label; the diagnostic label or classification may be a major purpose of assessment. Currently, classification is dictated by the *Diagnostic and Statistical Manual of Mental Disorders* (DSM-II, American Psychiatric Association, 1968), which presents a comprehensive framework with over 200 classifications and definitions for major categories. For example, the definition of *phobic neurosis* is:

This condition is characterized by intense fear of an object or situation which the patient consciously recognizes as no real danger to him. His apprehension may be experienced as faintness, fatigue, palpitations, perspiration, nausea, tremor, and even panic. Phobias are generally attributed to fears displaced to the phobic object or situation from some other object of which the patient is unaware. A wide range of phobias has been described. (p. 40)

This definition relies upon the psychodynamic notion of unconscious displaced (symbolic) fear. The DSM-II nomenclature for mental disorders uses both trait and psychodynamic concepts within the context of a medical-disease model (Ullmann & Krasner, 1969). The categories are clearly regarded as disorders or illnesses, conveying the notion that the behavior is "abnormal" or "pathological" and is due to an underlying cause analogous to a "germ theory" for physical illness.

Uses of Behavioral Assessment

Behavioral assessment procedures provide case formulations that focus on specific current maintaining or controlling factors, with terms or concepts such as *eliciting stimuli, reinforcing events,* or *escape-avoidance patterns.* In behavioral assessment, the S-O-R-K-C functional analysis yields a behavioral formulation that is also used to design and guide intervention procedures.

Behavioral assessment is a continuing and integral part of the therapy process (Kanfer & Phillips, 1970). It guides treatment from initial contact with the patient to termination and follow-up. This is apparent if we examine a systematic framework for guiding the process of behavioral assessment:

1. Problem identification
2. Measurement and functional analysis
3. Matching treatment to client
4. Assessment of ongoing therapy
5. Evaluation of therapy

Thus, a basic characteristic of behavioral assessment is that it is not limited to a pretreatment phase, but rather is an integral part of the therapy process itself. A detailed description of each of these stages will be presented in Chapter 2; specific use of the framework will be illustrated throughout the chapters on applications.

Methods of Assessment

Therapists, regardless of orientation, rarely rely upon one source of information; rather, multiple sources of data are usually sought to develop a working model or case formulation for treatment. In both traditional and behavioral assessment, the methods usually include: (1) the interview; (2) observation; and (3) psychological testing.

The Interview

Both traditional and behavioral assessment rely heavily on the interview. During interview sessions, the therapist's warmth and genuineness and the establishment of a positive therapeutic relationship are recognized as important by both traditional and behavioral therapists. The interview provides a rich source of information on the nature of the presenting problem and the patient's current functioning, history, and background.

Behavior therapists and traditional therapists differ in the importance placed on past versus present experiences. An emphasis on the patient's history is typical of traditional assessment. The interviewer generates hypotheses about the patient, based on this historical information. Attitudes the patient presently holds are seen as relating to attitudes the patient held in the past toward significant others. For example, an unassertive male patient who has major difficulties talking to women may be asked many questions regarding his childhood relationship with his mother.

In contrast, the behavioral clinician emphasizes the patient's present functioning. Current patterns of responding, antecedents, and consequences are focused upon. The behavioral interview starts with general information-gathering. This permits the client to describe the problem in his own style and the therapist to collect background information. Later, the behavioral interview is used to operationalize problems and to identify probable controlling variables.

To illustrate this process, consider the example of the unassertive male who has difficulties talking with women. The behavioral interviewer might initially invite the client to describe the problem in general terms. Later he might gather detailed descriptions of recent occurrences of the problem. Questions may focus on: (1) stimulus variables, such as age, attractiveness, or assertiveness of the women; and (2) consequences of the interactions, such as the client's embarrassment or escape, or negative reactions by the women. History is discussed and seen as important only as it relates to the present. Past social anxieties and skills are reviewed and linked to present assets and deficits.

Observation

Traditional assessment emphasizes careful observation of the client's behavior in the therapy setting itself. The patient's actions during sessions are seen as signs of underlying dynamics or traits. In addition, certain behaviors are interpreted as representing critical aspects of the psychodynamic therapy process, such as resistance or transference.

Playroom observations are often used in the traditional assessment of children. The playroom setting provides an opportunity to observe the child's behavior in an unstructured situation. The child's behaviors, toy selections, and manner of play are viewed as signs of underlying factors responsible for the child's problems. Although a wealth of behavioral data is available in such an evaluation, the child's behavior is rarely viewed as significant in its own right. For example, Simmons (1969) describes the behavior of a 7-year-old child who was referred for aggressive behavior at home and at school:

> He rushed to the playroom the instant he was invited. He quickly took the guns and shot wildly around the room with vivid sound effects and descriptive phrases such as "I got em! He's dead! You dirty Jap!" He tried to shoot the examiner. . . . (p. 11)

The observer concluded that this child was acutely anxious and that overactivity represented a method of dealing with his anxiety. It is likely that such a conclusion would direct the therapist away from dealing directly with aggression in the home and school. Therapy would likely focus on the alleviation of the child's presumed underlying conflicts.

Traditional assessment rarely emphasizes direct observation of the client in his natural environment. Rather, it is assumed that behavior in the office or playroom is representative of traits or dynamics that cause similar behavior in the home, school, or community.

Behavioral assessment, on the other hand, emphasizes observation of clients in a variety of natural settings. Observation of the patient in the therapist's office is not overlooked, but rather is viewed as a sample of behavior in a very specific setting. These samples provide the therapist with hypotheses regarding interpersonal style and other behaviors which are then investigated as to their generality in other settings. Observations in a variety of settings reveal the situational factors controlling behavior.

Behavioral assessment focuses on behaviors that can be observed and measured. For instance, one cannot observe a "poor self-image" directly, although this is a common complaint. But this complaint may be defined operationally by a client by his or her: (1) dating frequency; (2) performance at school or at work; and (3) frequency of negative self-statements. These components are all potentially detectable and measurable.

A variety of observational strategies are used by behavioral clinicians. These include self-observations, naturalistic or in vivo observations, and laboratory observations. These techniques are discussed in detail in chapters 2 through 6.

Psychological Testing

Clients referred for assessment are often given a battery of tests. Psychological test instruments traditionally have been used for purposes of intellectual, educational, neuropsychological, and personality assessment.

Intellectual, educational, and neuropsychological evaluations provide valuable information in understanding a client's strengths and weaknesses. The reliability, validity, and utility of these psychological tests are generally considered to be quite high (Mischel, 1968). They primarily sample performance and are valuable tools for behavior therapists as well as traditionally oriented therapists. They include tests of intelligence such as the WAIS or WISC, the Halstead-Reitan battery for assessing organic brain dysfunction, and learning disability tests.

The major difference between the traditional and behavioral approaches lies in their view of the utility of personality tests: Traditional clinicians favor personality testing while behavioral clinicians oppose it. Personality tests can take the form of projective or objective test instruments. In *projective tests,* a set of ambiguous stimuli is presented and the client is asked to respond. For example, the client may be shown various inkblots and asked what he sees in them (the Rorschach test), or he may be presented with a picture and asked to tell a story about it (the Thematic Apperception Test). The projective hypothesis underpins the use of such tests. The *projective hypothesis* asserts that responses to ambiguous material are signs revealing personality structure and dynamics. In *objective tests,* the client is presented with a series of written questions that typically require a response of "true" or "false." Scoring of objective tests is standardized and items are often grouped in scales which comprise a comprehensive personality profile. The Minnesota Multiphasic Personality Inventory (MMPI) and the California Psychological Inventory (CPI) are popular examples of objective personality tests. The objective tests are based upon traditional test construction theory.

Traditional clinicians favor personality testing because it yields hypothetical constructs that are presumed to have direct bearing on treatment. Since behavioral clinicians do not deal with ego defenses and repressed urges, they do not need personality tests designed to assess such variables.

Behavior therapists sometimes administer certain objective personality tests. In such cases, the tests are viewed as one limited source of data to supplement a comprehensive behavioral assessment. For example, the Depression Scale of the MMPI may provide a behaviorally oriented therapist with a self-report measure of depression to be used in conjunction with observations of the patient in a psychiatric ward. Behavior therapists, nevertheless, often do rely upon some paper-and-pencil psychological tests. The tests that they use primarily are intended to provide descriptions of behavior, antecedents, or consequences. This stands in marked contrast to the more global descriptions provided by traditional tests.

Strengths and Weaknesses of Traditional Assessment and Behavioral Assessment

Strengths of Traditional Assessment

Information gathered through traditional assessment procedures may guide the therapist in identifying content to be explored in therapy. This information aids the traditional therapist in dealing with clients' unresolved conflicts and gives him insight into relationships between feelings and behavior.

Traditional assessment provides the professional community with a shared language or terminology for describing clinical issues. Some traditional test instruments also conserve professional time. Many of these tests offer a distinct advantage in providing normative data by which to interpret test results.

Researchers have applied sophisticated statistical procedures to trait measures in attempts to discover the "structure of psychopathology" in the form of empirically derived factors (e.g., Lorr, Klett, & McNair, 1963; Eysenck, 1961). Although these derived systems have not yet gained acceptance by the clinical community, it has been suggested that they are potentially useful for research and therapy (Eysenck, 1973).

Weaknesses of Traditional Assessment

A major weakness of traditional assessment is that the basic assumptions underlying the assessment practices have not been adequately supported by personality research (Mischel, 1968). While there is little supportive evidence, there is much data to disconfirm the assumptions that test behaviors are valid signs or symbols of global and stable personality characteristics that determine behavior across settings. A second problem is the questionable reliability of assessment methods such as projective tests and the diagnostic classification system (Schmidt & Fonda, 1956).

The link between traditional assessment and treatment is a weak one. Diagnostic classification often leads to differential treatment on only a gross level (e.g., inpatient versus outpatient treatment). Similar treatment procedures are often used irrespective of assessment information (e.g., Daily, 1953).

Diagnostic labels generated by traditional assessment practices often have stigmatizing consequences (Rosenhan, 1973; Szasz, 1961). Critics point out that health professions and the community often respond to a diagnostic label in a vicious manner that persists after the "symptoms" that initially led to the label are no longer present. Such trait labels often contribute to the development or maintenance of psychological problems (Kopel & Arkowitz, 1975; Ullmann & Krasner, 1969).

Strengths of Behavioral Assessment

Behavioral assessment is committed to empiricism. Validity is increased by the use of samples rather than signs; specificity of analysis rather than global measures; and the strategy of empirically discovering functional relationships rather than relying upon a priori theoretical assumptions.

In addition to its emphasis on empiricism, behavioral assessment possesses clinical utility. The operational definitions, functional analyses, and measurement procedures lead to specific targets for change and guide the selection of treatment procedures. The therapist interphases assessment and therapy throughout treatment, using the information gathered for decision-making.

Finally, assessment is used to determine when specific treatment goals have been met and to evaluate outcome. Thus, assessment is closely linked to treatment across the entire assessment-therapy process. In summary, behavior therapy is guided more by measurable effects and less by intuition or assumption.

Weaknesses of Behavioral Assessment

One weakness of behavioral assessment stems from its reliance, at times, on verbal descriptions of behavioral events. Procedures such as interviews and paper-and-pencil tests violate the basic principle of direct sampling. If the clinician relies solely on these methods, the spirit of behavioral assessment is lost.

A second problem is the difficulty of collecting representative samples of behavior (Wiggins, 1973). The duration and extensiveness of sampling are important. Sampling should permit the natural variations of the environment to occur which control or maintain the problem and related daily functioning. Observational methods of assessment are often quite obtrusive and thus susceptible to measurement reactivity. The individual may behave differently than he usually does if he is aware that he is being observed or if he is observing his own behavior (see, e.g., Kopel & Arkowitz, 1974). Other related problems may stem from observer and client expectations, reactions to the demand characteristics of the testing situation, and reliability. These problems are realities the behavioral clinician acknowledges and attempts to minimize. Fortunately, these concerns are becoming increasingly popular research areas in and of themselves (Johnson & Bolstad, 1973).

The lack of normative behavioral data is a recently recognized weakness of the behavioral approach. Behavior therapists have only recently begun to address the issues of normative behavioral data and their utility for interpreting the data collected on the individual (see, e.g., Johnson, Wahl, Martin, & Johanson, 1973).

The costs of behavioral assessment in terms of professional time, effort, and resources are another problem—there are practical difficulties in applying the principles of behavioral assessment. Complex observational coding systems have been devised and used primarily within the context of large research grant programs which have financial and human resources well beyond those of typical clinical settings. Sophisticated instrumentation for psychophysiological analysis is inaccessible to most clinicians.

Practical problems inherent in behavioral assessment can be overcome. The problems of limited resources, for example, can be minimized by simplifying assessment procedures or by employing alternative ones without unduly compromising the quality of the information collected. Chapters 3 through 7 provide many examples of this process.

Implications of Behavioral Assessment for the Practitioner

There are a number of specific implications for the practitioner who chooses to attempt to follow the behavioral system of assessment. Perhaps the most important implication arising from a systematic behavioral assessment approach pertains to the role of the practitioner. In behavioral assessment the therapist assumes the dual role of clinician and scientist. He sets out to: (1) identify specific problems and targets; (2) empirically discover the functional relationships between the problematic patterns and the controlling and maintaining factors; (3) rationally select treatment procedures on the basis of assessment; (4) continuously monitor progress and modify treatment as indicated by ongoing assessment; and (5) evaluate the outcome of treatment. In other words, the clinician's decisions and evaluations are based upon scientifically sound evidence, rather than intuition or a priori theoretical assumptions.

The behavioral clinician uses different assessment tools or procedures than does the traditional therapist. In addition to the use of a new pool of paper-and-pencil tests, the practitioner will find himself involved in an entirely new arena for assessment, namely, the patient's natural environment. The interviewer or test examiner's office is no longer the only setting for assessment. The practitioner may serve as his own observer, or he may train and use paraprofessionals to function as observers, recording samples of the client's behaviors in the very settings where these behaviors are problematic (e.g., school, home, social situations). Along with the need for trained observers comes the need to develop or use systems of gathering, coding, and presenting data.

There are direct costs that the practitioner must bear if behavioral strategies of assessment are followed. The costs involve the additional effort and time necessary for measurement following the principles of behavioral assessment that have been outlined. The costs are often balanced by more efficient treatment. It is our belief that a cost-benefit analysis favors the behavioral approach. In the final analysis the incremental utility (Mischel, 1968) gained by the use of behavioral assessment far outweighs the costs.

CHAPTER 2

A Procedural Framework for Behavioral Assessment

The clinician who wishes to translate principles of behavioral assessment into practice faces several key questions. First, what are the actual stages involved in behavioral assessment? Second, what tasks must be accomplished at each of these stages? Third, what are the procedures used to accomplish these tasks? The present chapter aims to answer these questions. A procedural framework for behavioral assessment is presented. This framework is general enough to apply to varied populations and clinical problems. Later chapters will consider the specifics required for particular patient populations.

Behavioral assessment involves five general stages: (1) problem identification; (2) measurement and functional analysis; (3) matching treatment to client; (4) assessment of ongoing therapy; and (5) evaluation of therapy. These stages serve as a guide. They orient the practicing clinician through the process of behavioral assessment. Although the stages are presented here as distinct and sequential, in practice the overlap is considerable. Assessment may involve more than one stage at a given time. Thus, this framework is offered as a heuristic guide rather than a technician's manual.

Problem Identification

Problem identification is the first step in behavioral assessment. The major tasks facing the clinician at this point are: (1) pinpointing presenting problems; (2) determining response characteristics; (3) obtaining a history of the problem; (4) identifying probable controlling variables; and (5) selecting tentative targets for modification.

Pinpointing Presenting Problems

Assessment typically begins with an interview that provides the client with an opportunity to describe the presenting problem. The goal is to operationalize presenting complaints in behavioral terms. For example, vague notions of "feeling depressed" might be redefined in several ways. Feeling depressed may mean that the client stays in the house for long periods of time; that he rarely does things he likes to do; or that he finds it difficult to sleep. Presenting complaints are defined in specific behavioral terms that are relevant and meaningful to the client.

Behavioral problems are described in terms of one or more of the motor, cognitive, or physiological response systems. For example, an attendant's complaint that a patient was "crazy" or "schizophrenic" may indicate that problems are occurring in the motor response system (the patient paces frantically and makes odd hand gestures), in the cognitive response system (the patient claims that he is controlled by demons), or in the physiological response system (the patient hyperventilates).

Determining Response Characteristics

Once presenting complaints are defined in terms of specific responses, the next task is to determine the characteristics of these responses. Responses can be categorized along the following dimensions: (1) frequency; (2) intensity or magnitude; (3) duration; and (4) appropriateness.

Frequency estimates are often inaccurate. Initial descriptions by the client often suggest that a presenting problem occurs almost continuously. For example, highly anxious clients often state that they experience anxiety attacks "all the time." Upon questioning, however, the client may specify certain time periods or situations, such as weekends or parties, in which anxiety is either relatively low or high. Later, data collection on the frequency of problematic behaviors provides the clinician with more detailed information to either substantiate or refute the verbal self-reports.

Intensity is a second key dimension of behavior. For certain classes of behavior, the intensity-magnitude dimension is particularly influential in defining the pattern as problematic. For example, physical aggression is commonly observed in children. The intensity of hitting (and thus the degree of pain and damage inflicted) obviously influences the perceived severity of this problem.

Duration is a third important dimension of behavior. Duration refers to how long a behavior lasts. For example, a duration measure of a child's bedtime crying is the number of minutes he cries before he falls asleep. The duration of a behavior may be the primary dimension responsible for viewing the behavior as problematic.

Appropriateness. Notions of the frequency, intensity, and duration of a response must be viewed within the context of social norms and situational factors. The dimension of appropriateness of the behavior takes this social and situational context into account in an integrated evaluation of response characteristics.

Behaviors are classified as problematic because they are excesses or deficits (Kanfer & Saslow, 1965). Behavioral excesses are problems in which the response strength (in terms of frequency, intensity, or duration) is relatively too great. Clinical examples of behavioral excesses include compulsive behaviors, children's disruptive behaviors, and exhibitionistic behaviors. Behavioral deficits are problems in which the appropriate or adaptive responses are too weak (in terms of frequency, intensity, or duration). Examples of behavioral deficits include elective mutism and social skill difficulties. The distinction between behavioral excesses and deficits has important implications for the choice of treatment strategies (Cautela & Upper, 1975).

Obtaining a History of the Problem

In behavioral assessment, historical information is collected and directly linked to the current and potential variables which control or influence the current problematic patterns. Information regarding the history of the problem is usually collected during the first interview. The development of the problem and its stability over time are particular areas of concern. Historical information related to onset and prior adjustment provides useful information. The interviewer investigates thoroughly life events associated with the start of the problem, for example, a death in the patient's family, change in marital status, or loss of a job. Fluctuations in the severity of the problem and the events associated with these fluctuations are also explored. Such information provides the clinician with clues as to possible controlling variables. The client's past attempts to alleviate or cope with the problem are also examined for this purpose.

A comprehensive picture of problem history involves a review of many related areas. These include: (1) developmental history; (2) the current social-cultural-physical environment; (3) factors regarding drugs or medication; (4) intellectual capacities; and (5) physical limitations (Kanfer & Saslow, 1969). Life history questionnaires and interviews with family members are helpful aids in gathering this background information.

Identifying Probable Controlling Variables

Once the presenting problem(s) are clearly defined and the background information gathered, assessment focuses on the identification of probable controlling variables. The task here is to gather self-report information on the conditions which typically occur before the problematic behavior and the consequences that tend to follow it. A common first step in this process is to have the client describe a "typical day" in detail. Active questioning of the client yields a description of his daily activities from his awakening to his retiring. A review of the "typical day" indicates antecedents and consequences that may function to control behavior. A more detailed description and transcripts of this procedure appear in Chapter 3.

To further identify probable controlling variables, the behavioral interviewer may use more specific probes, such as: "Describe in detail the latest two occurrences of the problem."

Selecting Tentative Targets for Modification

The end result of this initial stage of behavioral assessment is the selection of tentative targets for assessment. Commonly, several targets can be specified. These targets are those which are most likely to lead to therapeutic change.

Major problems may arise if agreement between therapist and client on treatment goals is not established early. In behavioral assessment, an attempt is made to identify tentative targets for modification prior to behavioral measurement. These targets are discussed with the client and significant others in an open manner. When agreement as to goals and expectations for therapy is reached, behavioral assessment moves on to its second stage: measurement and functional analysis.

Measurement and Functional Analysis

The main purpose of the second stage of assessment is to collect more refined measurements of problematic behaviors and situational determinants. This stage provides opportunities to validate, modify, and adjust conceptualizations of the problems within the S-O-R-K-C formula prior to the start of treatment. There are two strategies which can be used to fulfill these purposes: *static analysis* and *functional analysis* (Ferster, 1965). A static analysis measures the characteristics of the behavior itself (i.e., frequency, intensity, and duration). In contrast, a functional analysis or "three-term contingency" (Bijou, Peterson, Harris, Allen, & Johnston, 1969) includes measurement of the antecedents, behavior, and consequences to assess controlling variables.

Measurement of Behavior

The first stage of assessment, problem identification, relied heavily on interview methods. Interviews are notoriously inaccurate sources of data. In the second stage of behavioral assessment, other methods of data collection are employed. These include questionnaires and observations which can be used for either static or functional analysis.

Questionnaires

A variety of specialized questionnaires and inventories have been designed to aid in problem measurement. Often these instruments measure the severity of the problem or identify associated situational factors. Different questionnaires are used with different patient populations (see chapters 3–7).

Observations

Observational procedures are fundamental to behavioral assessment because they can directly sample and record behaviors and associated environmental factors. Several forms of observation are used in behavioral assessment: (1) self-observation; (2) naturalistic observation; and (3) laboratory observation. These procedures are distinguished from each other by the individual who is performing the observation and recording or by the location and structure of the observational setting.

Self-observation. In self-observation, the client takes on the role of an observer. He records the incidence of a behavior and the surrounding circumstances. For example, the client can record the frequency and content of suicidal thoughts on a card throughout the day, noting the situational factors associated with these cognitions. Self-recordings are particularly useful for assessing cognitive responses that are not easily recorded by outside observers.

Naturalistic observation. Naturalistic observation involves the observation of the client's behavior by others in the natural setting. This strategy has been referred to as in vivo assessment. In some cases, checklists or elaborate behavioral coding systems are used by trained observers. More commonly, a parent, spouse, teacher, or paraprofessional is instructed to record the occurrence of a specific target behavior and perhaps the environmental conditions under which it occurs. For example, an aide might be asked to record the frequency of "paranoid statements" made by a patient, or a teacher might be requested to monitor the number of "call outs" made by a disruptive student.

Laboratory observation. Laboratory observation permits the direct sampling of behavior in a controlled setting. Role playing is one type of laboratory observation procedure. In role playing, the client is asked to carry out a specific role or task with the aid of the therapist or confederate and perhaps props. For example, males have been asked to simulate a meeting with a female stranger with the task of "getting to know one another" (see, e.g., Arkowitz, Lichtenstein, McGovern, & Hines, 1975). In other forms of laboratory observation, the client is exposed to a standardized set of stimuli (e.g., slides or videotapes) while responses are recorded. This strategy is particularly relevant to problems of behavioral excess. For example, sexual responses to inappropriate stimuli have been assessed in this manner by employing physiological measures of penile erection (Marks & Gelder, 1967; Rosen & Kopel, 1977).

Functional Analysis

Determining the specific factors that are functionally related to problematic behaviors is the most fundamentally important aspect of behavioral assessment. Functional analysis involves measuring the antecedents, consequences, and problematic behaviors in order to conceptualize the current controlling or maintaining variables. To illustrate this process, let us examine antecedents and consequences in more detail.

Antecedents are events or situational factors present immediately prior to the occurrence of problematic behaviors. Antecedents may include specific situational conditions or environmental events, the behaviors of others, and the individual's own behavior. Let us examine the case of an obese client who overeats primarily between 7 and 10 P.M., while watching television with a spouse who similarly overeats. A number of environmental antecedents may be functioning in this situation to control eating. Watching television in the living room is very likely an important situational factor. The spouse's eating in this situation probably also functions as stimulus for eating. The presence of high-calorie foods such as popcorn, potato chips, dip, etc., are physical stimuli that tend to elicit eating. The purchase of such foods and the initial act of eating them are self-generated stimuli that tend to lead to further eating. These latter stimuli often involve chains of behavior. Chains of behavior are composed of series of responses, each functioning as a stimulus for the next; they should not be overlooked because they often become targets for intervention.

Consequences are events or situational factors that follow the problematic behavior. They include changes in the physical environment, the situation, reactions of others, and alterations in the physical or physiological status of the client. For example, the child who has tantrums may do so because the parental attention which results is highly reinforcing. This is an example of positive reinforcement controlling behavior. A student may find it difficult to talk out in class because he stutters. In this case, punishment in the form of ridicule from peers may function to control behavior. A key consideration is: What does the individual obtain or avoid when he engages in the problematic pattern? For example, does the obese client overeat to avoid social or sexual situations that are threatening? Many behavioral patterns have additional reinforcers that accrue as the behavior is engaged in for long periods of time. For example, the diagnosed schizophrenic remaining in a Veterans Administration Hospital receives considerably more money than he would if he were discharged. Although this contingency may have had no functional relationship to the maladaptive behaviors initially, it may currently be a factor in maintaining "sick" behaviors.

If the function of the problematic behaviors is not correctly determined, treatment procedures which merely decrease the behavior may precipitate new problematic patterns. For example, the alcoholic whose drinking serves to reduce anxieties or to escape from an intolerable marital situation may find himself with new maladaptive behaviors to serve the same function, if only the drinking is eliminated in treatment. "Symptom substitution" often results from inadequate assessment of the function of behaviors (Ullmann & Krasner, 1965).

Cognitive mediational responses are a special class of antecedents or consequences that function as critical links in a behavioral chain. The individual's causal attributions, labels for emotional states, and self-perceptions may contribute to the maintenance of clinical problems (Kopel & Arkowitz, 1975). A critical area to investigate is the role of covert self-instructions (i.e., thoughts). These thoughts often play an active part in the problematic pattern (Meichenbaum, 1976). For example, the client's "fearful" statements to himself may increase anticipatory anxiety to the degree that his social performance at a party is severely disrupted, or he may even choose to avoid the party. The sight of a spider may lead to irrational thoughts about poison bites that, in turn, lead to escape. Prior learning experiences are important determinants of mediational responses.

In summary, the end products of the measurement and functional analysis stage are: (1) baseline (pretreatment) measurement of target problematic behaviors; (2) measurement and conceptualization of the functional relationships between controlling or maintaining variables and problematic behaviors; and (3) indications of targets to modify within the S-O-R-K-C complex.

Matching Treatment to Client

In the third stage of assessment, the information collected is integrated to arrive at the selection of treatment procedures. The major tasks in this process include: (1) assessing the client's motivation; (2) assessing the client's skills and resources; (3) selecting treatment procedures; and (4) conducting the information-sharing conference.

Assessing the Client's Motivation

An important factor in deciding what types of treatment strategies to use is the client's motivation. A behavioral test of motivation is already available in the way the client has previously carried out assignments. In the measurement stage, the client is asked to actively participate in assessment procedures. He may be asked to complete questionnaires at home or self-record certain classes of behaviors during the day. The client's performance on these tasks provides a sample of his approach to tasks requiring self-initiation and independent performance. If these attempts were poor, the use of self-administered or self-control procedures in the client's natural environment appears contraindicated. Instead, behavioral strategies which involve greater degrees of therapist control may be better choices. These conclusions about motivation should be made with caution, since an individual may fail to successfully complete assessment assignments for reasons other than motivational ones. For example, the client may not comprehend the instructions, may be unfamiliar with self-disclosure on questionnaires, may lack the skills needed to fulfill the assignment. If these factors are relevant, they too are important information for making decisions about matching treatment to client.

Assessing the Client's Skills and Resources

A second major consideration in choosing treatment procedures is the availability of resources. What are the behavioral assets of the client, his skills, his areas of adequate and superior functioning? What are the currently available and potential reinforcing and aversive stimuli? What are the client's limitations (e.g., intellectual capacity, physical limitations)? What is the potential for change in his physical-social environment? Who is willing to cooperate during therapy? What are their skills and capabilities, and what are the potential effects of their participation?

Selecting Treatment Procedures

In selecting a procedure from a set of available therapeutic procedures which appear to have equal chance of success, we can offer some general guidelines. First, choose the procedure which will be the simplest, least effortful, and least disruptive to this individual's life situation. Second, choose the procedure which best fits into his style of perceiving and solving problems. Thus, if an individual has a history of striving for self-control (and is highly reinforced by accomplishing control or mastery), the selection of behavioral self-control procedures over equally efficacious alternative techniques is indicated.

The choice of treatment techniques is influenced by several factors: (1) empirical validation of treatment techniques; (2) the assessment information itself; (3) the therapist's familiarity with a given procedure; (4) the availability of certain procedures (e.g., access to biofeedback equipment); and (5) practical issues, such as ease of administering a procedure. Clearly, clinical training, experience, and judgment are also major factors in selecting a treatment technique.

Once a treatment technique is selected, specifics of the application are worked out. For example, a decision to use a positive reinforcement program entails several more refined decisions. What reinforcers should be used and at what amounts? What is a reasonable starting point for the stepwise shaping program and what size subsequent steps should be used? To further illustrate, the use of a procedure like systematic desensitization requires such decisions for refinements as: (1) hierarchy items; (2) the use of in vivo or imaginal stimuli; and (3) the client's abilities for relaxation and imagery. With self-control procedures, the degree of therapist monitoring needs to be assessed. Should there be telephone contacts between therapy sessions to check on self-administered applications? For some clients extensive shaping or structuring either is unnecessary or is perceived by them as aversive or childish.

Conducting the Information-Sharing Conference

The "sharing conference" provides the final step in matching treatment to client before the therapeutic regimen is implemented. The therapist provides a comprehensive review of assessment information to the client. The behavioral case formulation is presented in language understandable to the client. The basic assumptions of the behavioral approach (i.e., principles of social learning theory) are reviewed and contrasted with the medical model. Care is taken to outline the details of the treatment plans, linking them to the assessment formulations. The client is given full opportunity to question and discuss assessment conceptualizations, rationales for specific treatment strategies, and the treatment procedures themselves. This conference serves to educate the client in the behavioral approach and to establish mutually agreeable goals. Importantly, it provides a structure to assess the client's willingness to accept the therapist's formulations and to start specific treatment interventions. Furthermore, the sharing conference may facilitate a "set" for client participation that helps elicit and maintain client cooperation throughout therapy (cf. Davison, 1969). The topics covered in a sharing conference are presented in the outline below.

Topic Outline for Sharing Conference

1. Presenting problems: brief review.

2. Development of current problems: An historical review of problem development, highlighting significant incidents, events, patterns of maladaptive behavior, and problems other than presenting ones.

3. Functional analysis of current major problems: Major problem behaviors are tied to current antecedents and consequences. Therapist uses simple diagrams to illustrate conceptualization.

4. Basic assumptions of behavioral approach: Review of the following concepts: The client may see himself as "crazy" or "sick"; however, research demonstrates that environmental manipulations can produce behaviors in "normals" which are functionally similar to the client's. Thus, the client's behavior can be understood using the same principles as those used to explain other types of learning. Also, "sick" and "crazy" are societal labels, not something that, like a disease, is inherent in client. Changes in present behavior depend on learning new responses to environmental and internal antecedents. The efficiency of learning depends on the client's consistency and motivation in practicing new responses.

5. Outline of proposed therapeutic regimen.

Assessment of Ongoing Therapy

At the start of this phase of assessment, the administration of treatment procedures is implemented. One of the major strengths of the behavioral approach is that the assessment process extends into and interfaces with treatment. There are three major tasks in assessing the ongoing progress of therapy: checking the client's use of clinical procedures or techniques, monitoring the effectiveness of the procedures, and finally, modifying the treatment program.

Checking the Client's Use of Techniques

If a behavioral technique is inadequately applied, it will likely be ineffective. Two questions need to be answered: Is the client using the technique at all? and How well is he using it?

A common problem encountered with behavioral self-control procedures is that assignments are just not completed. For example, covert sensitization (Cautela, 1967) usually entails daily practice sessions. Without regular practice, covert sensitization can never be given an adequate test. Early detection of noncompliance is essential. It allows the therapist to work on facilitating practice sessions or to select a treatment strategy that does not rely upon home assignments.

A second problem occurs when treatment procedures are not applied appropriately. For example, a couple employing the squeeze technique (Masters & Johnson, 1970) in the treatment of premature ejaculation may be applying the squeeze too early, too late, or without sufficient pressure. In this case, early feedback from the clinician can easily resolve the problem. With more complex treatment procedures, the assessment of inadequate application may suggest the need for additional training.

There are a number of methods available for assessing the adequacy of intervention applications. One can simply ask the client during the office session to describe in detail how he carried out intervention procedures. The client may also be asked to telephone in a report during the week. This has the advantage of giving him more immediate feedback and encouragement from the therapist. Procedural checklists to be completed by the client immediately after the assignment can serve to supplement this report and remind him of necessary procedural steps (see, e.g., Kopel, 1975). If the assignment requires verbal statements (e.g., practicing problem-solving skills by a couple with marital problems), audiotapes of the procedures are excellent tools with which to assess the quality of the clients' application.

Monitoring the Effectiveness of Procedures

When assessment confirms that treatment procedures are being applied appropriately, the next issue is to determine their efficacy. Is the use of implosive therapy lowering the frequency of exhibitionism? Is the application of a contingent/management program by parents reducing the level of inappropriate tantrums? The behavior therapist does not wait until therapy is completed to assess the effects of treatment. Data on changes in problematic behavior are collected on an ongoing basis. This therapy-assessment process permits a rational judgment as to when the therapeutic goals are met and when therapy should be terminated. In general, the methods of assessment employed during measurement in Stage 2 (measurement and functional analysis) can be used for monitoring change.

Modifying the Treatment Program

If the data reveal insufficient changes in the targets following appropriate applications of intervention procedures, two questions arise: Do the currently employed procedures need to be modified? Do different techniques need to be employed? For example, the steps in a graded task assignment may be too large or, as progress is made in a systematic desensitization program, new hierarchy items may become relevant.

Modifications in treatment programs may be indicated for a number of reasons, including:

1. incorrect target selection
2. inadequate assessment of the controlling variables (i.e., inaccurate conceptualization)
3. a mismatch between treatment and client
4. inadequate application of the procedures
5. unpredicted reactions to aspects of the program (e.g., habituation to aversive stimuli)
6. ongoing changes in the environment.

The monitoring of ongoing therapy permits readjustments to these problems. This process ultimately reduces the risk of failure. Currently employed techniques may be modified (for example, the use of different back-up reinforcers in a token-economy program) or replaced by a totally new intervention strategy.

Evaluation of Therapy

Although the assessment of progress in therapy is ongoing, the clinician usually makes a final evaluation also. The major tasks are evaluation of the outcome and maintenance of therapeutic effects.

Evaluation of Outcome

If the clinician follows all stages of behavioral assessment, he is in an excellent position to evaluate therapy outcome. He has taken measurements of target problematic behaviors at every step of the way. Measurements taken during the baseline phase of assessment provide a pretreatment evaluation of the effectiveness of intervention procedures. Data collected on targets during the intervention stage are compared to baseline measures taken earlier. Changes obtained are evaluated relative to initial goals set with the patient. If some problems have not been successfully modified or if new goals arise, the therapy contract can be renegotiated and the assessment process can begin anew. In some cases, referrals are appropriate at this time. Formal evaluations of therapy outcome can be made using single-subject experimental designs. The two most popular designs are the A-B-A-B withdrawal design and the multiple-baseline design. The purpose of these designs is to demonstrate experimental control over target behaviors. The clinician is typically less concerned with such goals than is the researcher; nevertheless, these designs are useful for clinical purposes as well.

A-B-A-B Withdrawal Design

This design involves the following sequence: (1) baseline; (2) treatment procedure; (3) return to baseline; and (4) reinstatement of the treatment procedure. Figure 2.1 illustrates use of the A-B-A-B withdrawal design to assess the efficacy of a contingent verbal reinforcement (praise) program in reducing the classroom interruptions of an 8-year-old girl (Gordon, 1976). Praise was administered by the teacher for nonoccurrence of classroom interruptions. Interruptions occurred at a relatively high level during the Baseline$_1$ observation hours, yielding a mean of 8 interruptions per hour. During Intervention$_1$, the frequency of interruptions dropped to a mean of 4.5 per observation hour. On return to baseline (Baseline$_2$, withdrawal of intervention), the frequency of interruptions was above Baseline$_1$ levels, with a mean of about 14 interruptions per hour. Reapplication of the treatment program (Intervention Procedure$_2$) had the effect of again reducing frequency of interruptions. This pattern of change in the target behavior supports the conclusion that the treatment itself was responsible for therapeutic improvement.

It should be noted that after the first application of treatment procedures, rival hypotheses such as maturation or coinciding extraneous events could have accounted for the decrease in interruptions observed. Only by withdrawing and then reinstituting treatment could these alternative explanations be ruled out.

Although researchers have long acknowledged the use of the A-B-A-B withdrawal design, therapists rarely apply the procedure in clinical practice. Nevertheless, such demonstrations on a clinical level may have profound effects on patients being treated and may, in fact, facilitate the continued application of the treatment techniques after termination. A reversal of effects obtained in this design is often dramatic proof to the patient that what he is doing is responsible for his improvement. Such attributions help to mediate future use of the intervention procedures by the patient or change agent if needed.

The A-B-A-B withdrawal design requires the assumption that treatment effects can be reversed by withdrawing treatment procedures. However, when treatment procedures have been in effect for relatively long periods of time, this assumption often is not valid. In these cases, environmental contingencies or changes other than those which are applied begin to take over and maintain the behavioral changes. For a more detailed review of this issue, as well as other concerns, such as the ethical issues of reversal of effects, the reader is referred to Leitenberg (1973).

Figure 2.1 A-B-A-B Withdrawal Design

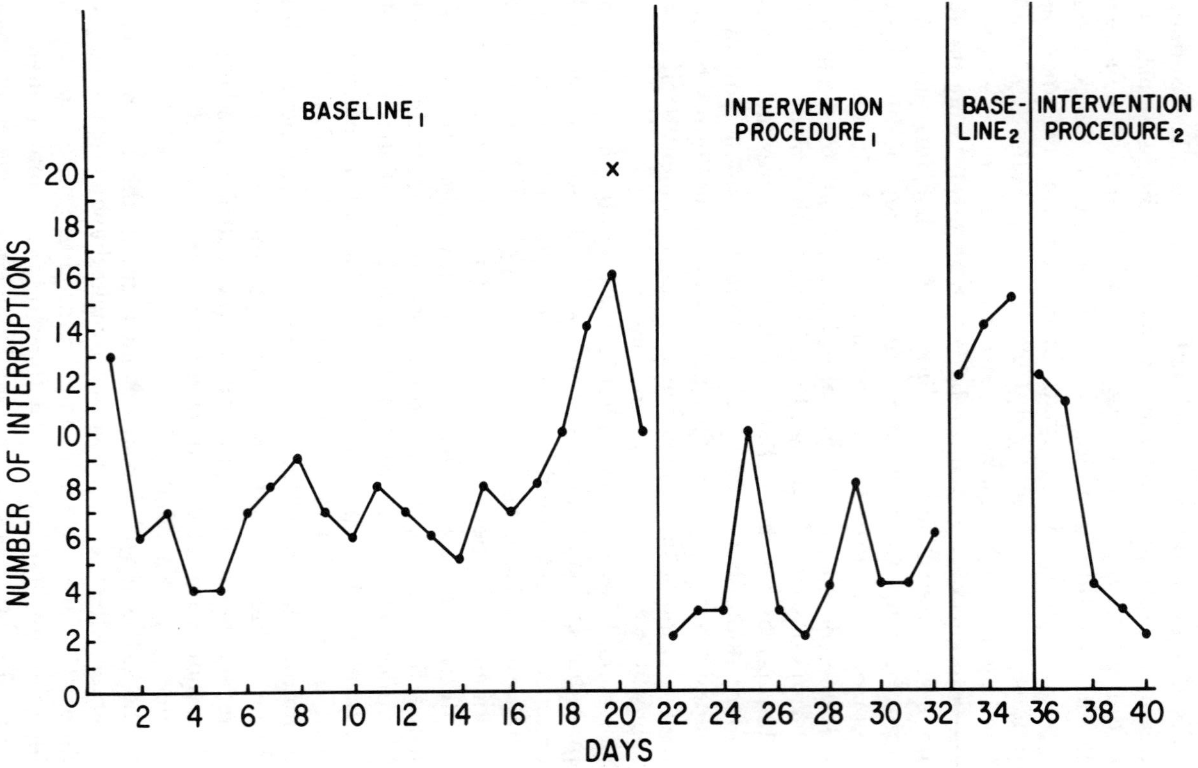

Reprinted from Gordon, 1976.

Figure 2.2 Hypothetical Example of the Multiple-Baseline Design

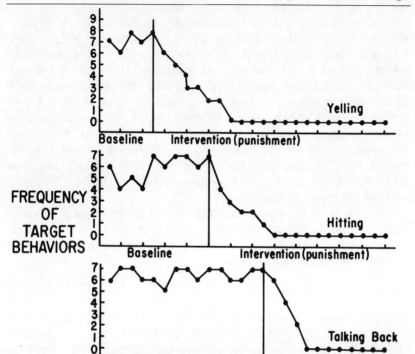

The Multiple-Baseline Design

The multiple-baseline design does not require the clinician to withdraw treatment. It relies upon systematic sequential application of a single treatment procedure across either: (1) several target behaviors; (2) several subjects; or (3) several situations. Figure 2.2 is an illustration of a hypothetical multiple-baseline design used to evaluate a contingent punishment program. The target behaviors to be modified include yelling, hitting, and talking back by a single child. Baseline data is collected for all three target behaviors and the intervention procedure is applied sequentially for yelling at Day 10, for hitting at Day 20, and for talking back at Day 30. The fact that each target behavior changed *when and only when* the procedure was applied supports the conclusion that improvement was caused by the treatment procedure. For a more detailed analysis of this design from a researcher's perspective, see Kazdin & Kopel, 1975.

In general, the multiple-baseline design has major advantages for clinicians over the use of the A-B-A-B withdrawal design. First, it avoids potential ethical problems inherent in withdrawal of treatment and reversal-of-treatment effects. Such issues are particularly relevant for certain classes of behavior, for example, self-destructive or aggressive behaviors. Second, the multiple-baseline design does not rely upon the continued reversibility of treatment effects. Rather, sequential applications of treatment procedures are used to rule out the influence of extraneous factors. Finally, the common clinical practice of working on one target at a time when multiple targets have been designated fits very well into the multiple-baseline design.

Single-subject experimental designs require continued data collection from baseline through completion of treatment. Thus, these designs can aid in the assessment of ongoing therapy as well as yielding a final evaluation.

Maintenance of Therapeutic Effects

Although issues of maintenance of therapeutic change are crucial, they are easily overlooked (Willems, 1974). Assessing the stability of behavior change requires the continuation of data collection.

One general strategy to use once goals have been met is for the clinician gradually to withdraw the structure of the therapy program while monitoring the durability of the behavior change. For example, the frequency of therapy sessions can be reduced to assess the stability of behavior change. Data collected during this period should provide some indication of the degree of maintenance to be expected when therapy is fully terminated.

A second method of assessing the maintenance prospects before therapy is completed is to ask the client to write out his own maintenance program (see, e.g., Lobitz & LoPiccolo, 1972). The program could include a description of: (1) old patterns and controlling variables; (2) new adaptive patterns; (3) danger signals indicating early or partial relapse; and (4) self-interventions, if needed. The client's write-up of this maintenance program provides the therapist with a useful tool. It demonstrates strengths and weaknesses in understanding and in other areas. If the write-up is returned to the client at termination, it may even facilitate the maintenance goals.

Follow-up sessions provide an important method of assessing

stability of behavior change once therapy has been terminated. In addition to potential benefits to clients, a follow-up assessment provides the clinician with information to evaluate his own treatment strategies. The practitioner who assumes the role of clinician-scientist uses feedback to alter his own behavior. It is this spirit of empiricism, applied to both the individual case and the therapist's cumulative performance, that uniquely defines behavioral assessment and therapy.

The next section of this book focuses on applications of behavioral assessment to different clinical populations. The five stages of behavioral assessment outlined in Chapter 2 could be applied to all clinical populations. In order to avoid undue repetition, each of the subsequent chapters will focus on the unique aspects of behavioral assessment relevant to that particular population. The reader, however, is reminded that comprehensive behavioral assessment does rely upon all five stages.

PART II
Applications

CHAPTER 3
Behavioral Assessment with Children

Little has been written regarding behavioral assessment with children. Only recently have there been attempts to provide a practical guide for those who work with children and their families (Gelfand & Hartmann, 1975; Miller, 1975; Tharp & Wetzel, 1969). In this chapter we consider some of the steps necessary for a clinician to conduct behavioral assessment with children. In keeping with the practical orientation of this book, we focus on the five stages of behavioral assessment (problem identification, measurement and functional analysis, matching treatment to client, assessment of ongoing therapy, and evaluation of therapy) as they apply to children.

Problem Identification

Parent Interview

A great deal has usually transpired before the behaviorally oriented clinician meets face-to-face with the child's parents. The parents, in all likelihood, have spoken to the child's classroom teacher, relatives, friends, their pediatrician, and perhaps even other therapists. Parents, therefore, often have their own expectations and ideas as to the nature of the problem and methods of treatment. It is essential that parental expectations be determined early in assessment.

The parents' preconceptions and expectations are more easily and naturally discussed when the therapist stays within their framework and language system for as long as is possible. The therapist-patient relationship is not likely to be enhanced if the therapist discusses the problems in terms of "discriminative and reinforcing stimuli" and the like. An example may help to illustrate this point. Below is a transcript of a novice behavior therapist conducting an initial parent interview.

> **PARENT:** Jack really lacks confidence, he's immature and doesn't relate to me.
>
> **THERAPIST:** Well, confidence and immaturity are concepts that make little sense. We have to become more objective.
>
> **PARENT:** I don't know what you mean.
>
> **THERAPIST:** What does he do that leads you to feel he lacks confidence?
>
> **PARENT:** Well, he's shy.
>
> **THERAPIST:** No! Shy is still too vague. We have to be able to count it. How can we count shy behavior?

The above is a pointed example of the behavior therapist not remaining long enough within the parent's framework. It would have been more useful had he used the parent's words (i.e., "confidence" and "immature") for a while longer. Also, to introduce the notion of record-keeping at this point is premature since the parents have not agreed to the concepts and use of a behavioral approach. The successful behavior therapist needs to know more than just the principles of learning; consultation and therapeutic skills are essential. Behavior therapists who ignore this element may have parents fail to keep records or, worse yet, drop out of treatment. The therapist then blames the parent for "sabotaging his excellent program." Perhaps greater success would accrue if "ownership" of the program were viewed more collectively. Lazarus (1971) points out the futility of therapists' forcing patients to accept a framework which is anathema to them. If parents subscribe to a certain approach, it is important to assess their cognitions regarding alternate models. A parent may be covertly saying: "Behavior modification is dehumanizing. We have to get at feelings to see real change." The beliefs people have and the internal sentences they say to themselves have a powerful effect on overt behavior. If such thoughts are ridiculed or ignored, cooperation and motivation are hampered.

Once the therapist feels that he has begun to establish a relationship and, when appropriate, that he has acknowledged the existence of other models of treatment, then data which have a more direct bearing on treatment can be collected. At this point, the consultant should attempt to identify the behavioral referents of the presenting problem. Gelfand & Hartmann (1975) have provided an outline to help the therapist to structure the initial interview so as to specify behavioral referents (see Table 3.1).

According to Mischel (1968), most laymen, as well as traditional therapists, are trait theorists in their descriptions of behavior.

Table 3.1
Initial Caretaker Interview

These are the basic questions to ask caretakers concerning the child's problem behavior:

1. *Specific description*
"Can you tell me what (*child's name*)'s problem seems to be?"
(If caretaker responds in generalities such as, "He is always grouchy," or that the child is rebellious, uncooperative, or overly shy, ask him to describe the behavior more explicitly.)
"What, exactly, does (*he or she*) do when (*he or she*) is acting this way? What kinds of things will (*he or she*) say?"

2. *Last incident*
"Could you tell me just what happened the last time you saw (*the child*) acting like this? What did you do?"

3. *Rate*
"How often does this behavior occur? About how many times a day (*or hour or week*) does it occur?"

4. *Changes in rate*
"Would you say this behavior is starting to happen more often, less often, or staying about the same?"

5. *Setting*
"In what situations does it occur? At home? At school? In public places or when (*the child*) is alone?"
(*If in public places*) "Who is usually with him? How do they respond?"
"At what times of day does this happen?"
"What else is (*the child*) likely to be doing at the time?"

6. *Antecedents*
"What usually has happened right before (*he or she*) does this? Does anything in particular seem to start this behavior?"

7. *Consequent events*
"What usually happens right afterward?"

8. *Modification attempts*
"What things have you tried to stop (*him or her*) from behaving this way?"
"How long did you try that?"
"How well did it work?"
"Have you ever tried anything else?"

Reprinted from Gelfand, D. and Hartman, D., *Child Behavior and Analysis*, Copyright, 1975, Pergamon Press, Ltd.

Parents appear to be notorious in this regard. Typical presenting complaints are that a child is aggressive, rebellious, hyperactive, lacks confidence, or has a poor self-image or a "mind of his own." Although they are far from precise, the therapist should adopt these labels and use them as his own during the initial stages of the interview. An example will help to illustrate this point.

> **MOTHER:** My child has such a poor self-image. He really doesn't like himself very much.
>
> **THERAPIST:** Now, let's see, Jack is seven. How long has he had a poor self-image?
>
> **MOTHER:** I'd say it goes back at least several years. I guess when he was about three.
>
> **THERAPIST:** So, prior to three, there were no problems with his self-image?
>
> **MOTHER:** No, none I could think of.
>
> **THERAPIST:** Can you recall if there were any significant changes in his life at that time?
>
> **MOTHER:** Well, I became pregnant and it was rough. I was sick and I guess I began to play with him less frequently.
>
> **THERAPIST:** So his poor self-image began at around the time you were having physical problems and out of necessity had to give him less attention.
>
> **MOTHER:** Yes, I hadn't thought of it like that before, but that makes sense.

At this point the therapist has little idea as to what kind of behavior the mother is referring. Nevertheless, by remaining within her language system, communication is facilitated. Once ease of communication is established, behavioral referents are sought. Every behavior therapist is aware that "a poor self-image" refers to a hypothetical construct. Often very simple questioning will reveal this information. The timing appears to be a critical element.

> **THERAPIST:** Well, you have mentioned Jack's poor self-image, could you tell me more about that?
>
> **MOTHER:** He's so hard to handle, he battles me every step of the way and I've read books that say a child who resists the way he does must have a poor self-image.
>
> **THERAPIST** (*takes note of the book the mother has read as something to come back to*): In what way does he resist?

> MOTHER: I can give you an example. I have to dress him every morning. He's seven but he won't dress himself. I've begged him, threatened him, bribed him.
>
> THERAPIST (*takes note of previous attempts to change the behavior. He will come back to this later*): Are there other instances of resistance?
>
> MOTHER: Well, he just doesn't listen to me. I'm always having to tell him to stop this or that, over and over. He doesn't do as he's told.

Here is a situation where the mother stated the problem as a poor self-image. The process of specifying behavioral referents has begun. With little prodding, the parent has indicated that failure to dress himself and noncompliance are at least two components of his "poor self-image." In most situations, this is not a difficult progression for parents, providing the therapist makes use of appropriate interviewing skills.

Once problems are specified, it is helpful to get the parents' estimate as to the problem's frequency or duration. The therapist could have the parents estimate whether dressing was a problem once a week or every day. Did the child fail to comply once a day, ten times a day, 50% of the time, or 75% of the time? These data are not necessarily viewed as completely accurate, but they do provide a picture of the parents' perception of the problem.

Since a truism of behavior modification is that behavior is a function of the environment, the therapist begins to gather information regarding probable controlling variables. We emphasize the probabilistic nature of the variables here to underscore the hypothesis-generating and -confirming approach of behavioral assessment. One approach we have found very useful in identifying behavioral referents, estimating frequency, and identifying controlling variables is to ask the parents to describe a typical day in their family from the time the first person is up in the morning to the time everyone retires. This single question has been found to yield a rich source of information, often requiring an entire session. An example follows:

> THERAPIST: Could you describe for me a typical day in your family? I'd like you to be as specific as possible.
>
> MOTHER: Well, my husband is up before anyone and is usually out of the house before the kids even get up. Then, I get up at about 7:00 and yell to the kids to get up.
>
> THERAPIST: What is the sleeping arrangement for the children?

MOTHER: Do you mean, "Do they sleep in the same room or separate rooms"?

THERAPIST: Yes.

MOTHER: Oh! They each have their own room. This is where the problems start. I just can't get him up.

THERAPIST: Can you tell me what typically occurs?

MOTHER: Well, after yelling to him to get up, I go into his room after I have taken care of myself and he is usually still in his pajamas and is either under the covers or is playing with some toys in his room. On occasion he is downstairs in front of the T.V.

THERAPIST: What do you do at this point?

MOTHER: Well, I wind up dressing him because it's too much of a hassle the other way. I know you're going to tell me not to do it but I can't let him miss school and I have to get to work too.

THERAPIST: How does that work out when you dress him?

MOTHER: Sometimes, when he's in a good mood, he cooperates and at other times he fights me every step of the way.

THERAPIST: Do you do or say anything to him on those days when he does cooperate?

MOTHER: Oh. I have told him how happy I am but then this just seems to make it worse.

THERAPIST: How do you handle it when he doesn't cooperate?

MOTHER: I usually have to raise my voice and he stops. Sometimes I have to hit him.

THERAPIST: How often would you estimate that he cooperates each morning by letting you dress him?

MOTHER: Gee, it seems like never. I would say maybe three-quarters of the time he gives me a battle. But you know its a funny thing; we never have this problem on weekends when he wants to go out and play. He gets dressed by himself, no problem at all.

THERAPIST: How do you view that?

MOTHER: Maybe it's me. I think he can do it when he really wants to.

THERAPIST: How often does he dress himself?

MOTHER: Usually only on weekends but every once in awhile he surprises me.

THERAPIST: What happens after he is dressed?

The interview proceeds in this fashion through the rest of the day. Information is readily available as to problematic areas. Problems often center around the following morning activities: getting

Table 3.2
Checklist for Problematic Areas

Name of child _____ Date_____
Completed by _____

Place a checkmark (✔) in the appropriate space if you have observed any difficulty(ies) within the past two weeks.

Morning

_____ Getting up
_____ Dressing
_____ Breakfast
_____ Sibling/peer relationships
_____ Leisure time
_____ Leaving for school

After School

_____ Return home from school
_____ Chores
_____ Homework
_____ Sibling/peer relationships
_____ Leisure time
_____ Dinner

After Dinner

_____ Chores
_____ Homework
_____ Sibling/peer relationships
_____ Leisure time
_____ Bedtime

up, dressing, eating, and leaving the house for school. Since the children are at school for a large part of the day, the interview temporarily bypasses school problems and focuses on the time between the child's return home from school and dinner. Typical problem areas occurring at this time of day center around: the child's return home, household chores, peer or sibling relationships, homework, and dinner. After-dinner problems typically center on play activities between peers or siblings, and bedtime difficulties. The bedtime period should be explored in detail as it often reveals unmentioned problems (e.g., enuresis) and also provides a picture of the presence or absence of positive feelings between family members. A "ritual" of bedtime stories and closeness may or may not be present. Table 3.2 presents a checklist the therapist can use to assess the range of problem areas. By administering this checklist at the end of treatment, a general index of improvement is easily obtained.

In addition to highlighting problem areas, the analysis of a typical day often crystallizes parental differences with regard to values, beliefs, and the nature of the problem. Skillful questioning which focuses on the social interaction between the child and his parents helps the therapist to identify probable controlling variables. An example follows:

> **MOTHER:** I can't seem to get him to understand that when I say "no" I mean it.
>
> **THERAPIST:** Well, when you say "no," what does your son do?
>
> **MOTHER:** Well, first he will start to whine and say, "Everyone else does it."
>
> **THERAPIST:** What do you do then?
>
> **MOTHER:** I start to feel guilty because I work and can't give him all the attention I should, so I try to explain to him why he can't have whatever it is.
>
> **THERAPIST:** What happens after you have explained that to him?
>
> **MOTHER:** Oh, it's not "after," I don't even get that far. As soon as I try talking to him in a pleasant, calm voice he starts to say I'm unfair and begins to cry or tantrum.
>
> **THERAPIST:** What happens when he starts to tantrum?
>
> **MOTHER:** Well, sometimes I give in and sometimes I try to reason with him and sometimes, when I've had a hard day, I let him have it. I know I should be more consistent but it's so hard.

The therapist should resist the impulse to comment on what the parent is doing, right or wrong. Experience shows that "debates" between therapist and parents at too early a stage, before a solid working relationship has been established, are counterproductive. Rather, the therapist should begin to develop hypotheses regarding possible controlling variables. In the above illustration, it is reasonable to speculate that the mother is reinforcing the child's noncompliance by her attention. When she tries to set limits, her son's deviant behavior, which is reinforced by the mother's giving in, escalates. The reciprocal nature of this coercive pattern occurs because the mother is reinforced for giving in by the termination of an unpleasant scene (Patterson & Reid, 1970). This explanation is consistent with the social learning approach put forth by Patterson (1971).

As has been pointed out by several behavior therapists (e.g.,

Lazarus, 1971; M.J. Mahoney, 1974), controlling variables may be covert as well as overt. The domain of behavior therapy has been recently extended to include cognitive factors (see, e.g., Kopel & Arkowitz, 1975). This trend is influenced by both empirical research (Bandura, 1969; M.J. Mahoney, 1974) and direct clinical investigation (Lazarus, 1971). Behavior therapists working with parents must begin to assess parents' cognitions that may influence their behavior. If a mother is continually saying "I am a bad mother for working and therefore I have to give in to all my child's demands in order to make it up to him," it would seem that therapy would be incomplete and susceptible to short-lived success unless her cognitions are modified as well. (For a more detailed explanation of how behavior therapists restructure cognitions, see Lazarus, 1971, and Ellis, 1962.)

Although the therapist is beginning to develop hypotheses about what is controlling the behavior, some notions about the initial causes of the problem are also being developed. Typically in behavior therapy, less attention is paid to historical antecedents than to current, maintaining factors (Bandura, 1969). A potential problem exists in that parents may feel that the initiating causes are what is most important. Because the therapist fails to attend to presumed causes, the parents may feel that they are receiving inadequate treatment and this may lead to poor cooperation and even to premature termination. Therefore, it is suggested that the therapist make some attempts to explore the parents' attribution as to the causal factors. Some parents may already believe that their behavior and their style of interacting with their child is a major factor contributing to the existence of the problem. Other parents, however, may believe that the problem lies solely within the school or within the child himself. The information on parents' perception of causal factors should help the therapist to personalize his approach (Lazarus, 1971). Of course, it may turn out that the parents are adamant that the only way their child will improve is if he has a therapist to talk to once a week and that there is really little that they can do to modify his behavior or their own. At this point, the best strategy is to make a referral to an appropriate therapist. Therapists who view their work with parents in too mechanical a fashion are simply conducting ineffective assessment and therapy (for example, if they merely hand the parents a primer on the applications of social learning theory [see, e.g., Patterson & Gullion, 1968] and meet the parents once a week to check on the record-keeping).

In order to conduct a behavioral analysis of a child's behavior, we must examine the parents' behavior. As we have seen, relevant parent behaviors include not only daily behavioral interactions, but also parents' cognitions regarding parenting. If we acknowledge that, in order to change their children's behavior, parents must change their own behavior, all the factors of social influence become relevant. Parents must make some difficult changes in their behavior. The therapist's positive working relationship with the child's parents is important in facilitating these changes. Establishing and maintaining this relationship requires considerable clinical skill and sensitivity. In summary, the parent interview, as the first step in the assessment procedure, requires all the clinical skills associated with being an effective therapist.

As mentioned in Chapter 1, there is an accumulation of research that calls into question the validity and utility of trait and state theories (Mischel, 1968). This information has a direct bearing on the assumptions made by the therapist. One assumption that is *not* made is that the way in which the child functions at home is a good predictor of the way he functions at school. A more accurate statement is that the greater the difference in the two settings, the less the predictive validity. Therefore, it is necessary for the therapist to question the parents about the child's functioning in school. This applies not only to academic areas and social behavior but also to the parents' attitudes toward the school situation. In addition to a detailed review of the child's present behavior, an historical account of his school career is valuable. The information provided by the parents about their child's school behavior is viewed with an "I'm from Missouri—Show me" attitude. When possible, contact with the school is scheduled to obtain first-hand information.

Interview with School Personnel

One of the most sensitive areas in the behavioral assessment of children is the therapist's entry into the school system. Typically, the therapist needs to do an assessment of the child in the school setting when the complaint originates from the parents' concern for their child's overall adjustment or when it originates from within the school system itself. Over and over again, we have seen classroom management programs which look excellent on paper but are never implemented or, when they are, the final refrain is: "The teacher sabotaged the program." Knowledge of behavioral principles is necessary but not sufficient for the effective implementation of behavioral intervention.

Behavior therapists need to consider social and political issues of systems consultation (Mannino, MacLennan, & Shore, 1975; Repucci & Saunders, 1974). Questions to be considered include the following:

1. Did the classroom teacher initiate the referral procedure? Does she view the child as a problem, or is she surprised that a professional contact was made?
2. Has the teacher worked with outside consultants in the past? What were these experiences like? Were they beneficial?
3. How does the teacher feel about the use of behavior modification? Has she had any didactic or practical experience in this area?
4. What are the teacher's goals for this child? Is she interested in helping him to improve so he can be maintained in her class, or is she intent on collecting information so that she can get the child out of her class?
5. What is the role of other school personnel in the case (the principal; teachers of special subjects, e.g., gym and art; the school psychologist; social workers)?
6. If other school personnel are able to exert influence on the classroom teacher, how do they feel about outside consultation? How do they feel about the use of behavior modification in the classroom?

These questions raise issues that merit attention and consideration for successful therapeutic intervention in the school environment. The answers provided help to guide the therapist as to which school personnel need to be involved in the assessment phase of treatment. An example may help to illustrate this point. One of us recently saw the parents of a 7-year-old boy who had been referred by a school social worker. Following the parent interview, permission was obtained to contact the school. A telephone call to the social worker revealed that she did not have the time to become involved in the treatment planning, but wanted to be kept informed as to future developments. The child's teacher was aware of behavioral problems in the classroom and was interested in working closely with the therapist. In the child's school there was also a learning disability examiner who viewed his specialty as behavior modification. This examiner had a strong interest in becoming involved in any behavioral treatment program with the child. It would have been extremely unwise to bypass him at this stage. Interviews were scheduled, as a result, with the classroom teacher and the learning disability specialist.

This example illustrates two points: the need to assess the intended involvement of the referring agent, and the need to assess the feelings of school personnel who view the therapist's activities within their domain. Failure to involve significant change agents from the school system early in the assessment process can seriously impede future cooperation.

The guidelines suggested for the parent interview apply for interviewing the school personnel as well. In all cases, interviewing skills and the ability to establish working relationships influence assessment. Many students and even experienced clinicians seem surprised to learn that this is a necessary step in the process of behavioral assessment and therapy.

Checklists and Questionnaires

A characteristic of behavioral assessment is the use of multiple sources of data collection. In addition to the verbal interview, parents and teachers are often asked to complete various behaviorally oriented checklists and questionnaires. The information gathered from these instruments is useful in providing the therapist with information omitted in the interview. In addition, contradictory or confirming data is obtained. Finally, a referent group is often provided when the checklist or questionnaire has been used with a large population of cases.

Parent Questionnaires

Prior to the first session, it is helpful to have the parents complete a questionnaire which provides biographical and demographic data as well as indicating the way in which they view the problem. The Family Questionnaire has been useful for these purposes (see Appendix A).

The therapist has many paper-and-pencil instruments from which to choose. These instruments are used to assess the parents' perceptions of their child's behavior as well as to assess the overall level of problem severity. Ideally, the test chosen should meet all the criteria essential for adequate test construction (Cronbach, 1960). The Louisville Behavior Checklist (LBCL) developed by Miller (1967) is especially suitable for this population. The LBCL has the advantages of being well constructed and providing the assessor with the options of quick scanning or more formal scoring. The LBCL has been standardized for boys and girls aged three to 13 and yields scores on three broad factors and eight narrow factors. The three broad factors are Aggression (AG), Inhibition (IN), and Learning Disability (LD). The eight narrow factors are Infantile

Aggression (IA), Hyperactivity (HA), Antisocial Behavior (AS), Social Withdrawal (SW), Sensitivity (SN), Fear (FR), Academic Deficit (AD), and Immaturity (IM). In addition, an overall score provides information as to the Total Disability (TD). The LBCL provides the clinician with information regarding parents' views of their child and differences in their views. It also alerts the therapist to problems not mentioned during the initial interview and provides for a comparison with a normal population. It can be used as an assessment of outcome. We have been using the LBCL routinely with parents who have participated in a behaviorally oriented parent group (Gordon, Lerner, & Keefe, in press).

In any type of therapeutic work with children that involves both parents, it seems appropriate to assess the quality of the marital relationship. The manner in which the parents relate to each other may be in need of modification prior to any intervention that focuses more directly on the child. In addition to discussing marital problems during the interview, each parent is also asked to complete a questionnaire that provides information about the marriage. For this purpose, we use the Locke-Wallace (Locke & Wallace, 1959) Marital Adjustment Scale (MAS). (The MAS is reproduced in Chapter 5.) If, in addition to deviant behavior on the child's part, severe marital discord seems to be present, the treatment plan needs to be uniquely tailored to meet this situation. (See Patterson & Hops, 1972, and Stuart, 1969, for descriptions of marital behavior therapy.)

Teacher Questionnaires

The therapist's task of interviewing teachers is made difficult by the limited time available to conduct a leisurely interview. A supportive school system can arrange for the clinician to interview a teacher during a free period or after school; however, since time is often scarce, the use of paper-and-pencil questionnaires can be very helpful. Questionnaires, however, are not a substitute for a skillfully handled verbal interview.

In working with teachers, the Walker Problem Behavior Identification Checklist (WPBIC; Walker, 1970) has been most helpful. The WPBIC is a 50-item weighted checklist which yields scores on factors of Acting-Out, Withdrawal, Distractibility, Disturbed Peer Relations, Immaturity, and a Total Deviant Score.

In addition to the WPBIC, a number of open-ended questionnaires have been developed. These are used to collect information on the teacher's view of the problem, its frequency, and assumed antecedents and consequences (see, e.g., Alper & White, 1971).

Problem Selection

Once the range of home and school problems has been clearly specified, the next step in assessment is to select a problem behavior in need of modification. There are several practical issues which need to be considered in problem selection. Because of the multidimensional approach used in problem identification, the behavioral clinician often uses several methods to identify problems, such as interviews and questionnaires. While this multidimensional approach has long been advocated (Eysenck, 1961), it does create difficulties. A major difficulty is that different methods of measuring the same problem often fail to produce similar results. A child who is rated high on a factor of aggression on a behavior checklist may be rated low on the same factor as a result of an interview with a parent or teacher. The therapist should not be surprised if discrepancies between teachers' and parents' information appear. Research in this area suggests that there is little consistency of behavior across environments. Walker, Hops, and Johnson (1975) concluded that children who exhibit high rates of deviant behavior in school do not necessarily show similar difficulties at home. Attempts to study the relationship between behaviors in the home and school as reflected in factor-analyzed questionnaires have yielded low correlations. The LBCL for females and its school counterpart, the School Behavior Checklist, have yielded correlations of .21 for the factor of Aggression and only .38 for Total Disability (Bloch, 1971). The correlations for males are slightly higher. These differences, when they occur, may be due to real differences in the child's behavior or to differences in perceptions by the raters, or to a combination of both. The implications of this problem for research on child behavior therapy have been discussed in an excellent article by O'Leary (1972). In practice, when such discrepancies between information sources occur, they are openly discussed and clarified with the mediators.

Another difficulty might arise if the parents' view of the severity of the problem is discrepant from that of the therapist. Parents may overestimate or underestimate the seriousness of the problem. The 3-year-old boy who wets his bed four nights a week is an example of the former, whereas the 9-year-old boy who coerces his mother to lie in bed with him each night until he falls asleep illustrates the latter. If such discrepant views are held, the therapist is often able to exert his position as an authority and experienced professional. Problem selection, then, not only requires basic knowledge of developmental and abnormal psychology, but also clinical skill in dealing with sensitive interpersonal relationships.

A third important issue to consider with regard to problem selection is whether or not the problem behaviors are part of a general response class. It has been our experience that noncompliance appears to be a class of behavior that when changed in the desired direction has profound positive effects on target behaviors not specifically treated. Problems such as thumbsucking, whining, and tantrums, may cease to occur when a child's noncompliance is reduced. This is information of which we would not expect parents to be aware. Sharing this information with parents is often helpful and may save countless hours of treatment.

A final issue is posed by the situation in which the therapist is faced with the dilemma of allowing parents to choose a problem that may be most relevant to them and yet is also the most difficult to change. The therapist may ask himself whether it is better to let the parents choose to work on a difficult problem, which is more likely to fail, or on an easy problem, which has a relatively good chance of success but may not be viewed as relevant. Eyberg and Johnson (1974) recently studied the relationship between the difficulty of the problem and treatment outcome. Families being seen for behavior modification were assigned either to a condition in which the problem that appeared to be the easiest was dealt with first, or to a condition in which the most difficult behavior problem was treated initially. A problem was considered easier to treat if it could be observed by the parents, occurred with a high frequency, was maintained by the parents' social interactions, and occurred at a specific time. The results indicated that there were no advantages in beginning treatment with an easier problem. In practice, we typically allow parents to make the decision about problem selection unless there are clear-cut contraindications. This appears to enhance their motivation and facilitates active participation in treatment planning and implementation.

Following problem selection but prior to intervention, it is helpful to discuss specific and objective longer-range goals. Failure to do so results in the "changing goal" phenomenon. Initially, parents may state that they would be pleased if their child would just stop wetting the bed. Upon the elimination of this problem, the parents may state they are not satisfied because their child still does not listen to them. With each accomplishment, the parents take change for granted and proceed to "up the ante." Frustration and disappointment are minimized if behavioral goals are specified in advance.

Measurement and Functional Analysis

After the identification and selection of the problem, the next step in behavioral assessment is the measurement of the problem. Essentially, there are two strategies which can be carried out. These strategies have been referred to as *functional analysis* and *static analysis* (Ferster, 1965). (See Chapter 2, p. 20 for a definition of each.) Functional and static analyses with children are usually carried out by means of direct observation by the mediators (parents, teachers, aides, babysitters) and the therapist. An awareness of the difficulties inherent in gathering and interpreting observational data is important; the clinician needs to be familiar with them so that he does not draw unfounded conclusions from the data. For a detailed discussion of the difficulties involved in using data from observations, the reader is referred to Johnson and Bolstad (1973).

Observations by the Mediator

Once the rationale for data collection has been fully explained and the therapist feels satisfied that the mediator has accepted this rationale, a data collection procedure is chosen. Since parents and teachers are not accustomed to record-keeping, it is important to make the first attempt successful. Success is maximized by selecting a simple observation procedure as well as by providing structure.

When the data from the parental interview regarding parent-child interaction is inconclusive, functional analysis over one or two days is necessary. To facilitate data collection according to the functional analysis or "three-term contingency" (antecedents, target behavior, and consequences), a 3-column observation sheet is used. Gelfand and Hartmann (1975) have suggested the following rules for such observations: Describe only the behavior or stimulus events that can actually be observed; record nonverbal as well as verbal behavior; note the approximate time a new behavior is initiated; and develop shorthand notations that allow for the recording of all actions.

Collecting data using this format not only provides information to the consultant, but also provides feedback to the parents as to the quality of interactions. For example, recently one of us saw the parents of two boys aged six and three. The parents were concerned about what they believed was extreme jealousy on the part of their older son whenever he was not the center of attention. The

Table 3.3
Results of a Mother's Three-Term Contingency Data Collection Procedure for One Week

Name: Benny
Date of Observation: 8/3/75–8/10/75

Setting Events: Mother, father, two children sitting in family room watching T.V. after dinner. Father reading newspaper.

What Happens Before	Behavior	What Happens After
8/3 6:50 P.M.: Entire family watching T.V., Marty sat next to me.	Benny said, "I was going to sit there," sucking thumb.	Lou began to sternly reprimand Benny. I made room for him and he sat next to me.
8/4 11:30 A.M.: Benny was showing his grandmother some new tricks he learned in gym; Marty began to imitate the tricks.	Benny pushed Marty, sucked thumb, frowned.	I asked Marty to sit down and wait until Benny was finished.
8/6 7:30 P.M.: Lou and the boys were playing basketball and he let Marty go first.	Benny frowned and ran into the house.	Lou followed Benny into the house and tried to explain that he he had to share.
8/8 4:30 P.M.: Benny came in from playing a game with Marty.	Benny threw a book at me and sucked his thumb.	I hit him several times and sent him to his room I went up to him 15 minutes later and we talked things out.
8/8 8:00 P.M.: I asked Benny and Marty a a question and Marty answered first.	Benny hit Marty.	I tried to talk to both of them for 5 minutes.

parent interview suggested that no severe problems existed in other areas of the child's life or in the marital relationship. This tentative conclusion received support from data gathered by the Louisville Behavior Checklist, the Locke-Wallace Marital Adjustment Scale, and a telephone conversation with the child's teacher. Since the parents stated that the problem was not too frequent in occurrence, they were instructed in the use of the "three-term contingency" data collection procedure and they agreed to try it for one week. The results of the data collection are shown in Table 3.3.

At the subsequent meeting, both parents reported a "revelation" with regard to the problem. They reported that they became aware of how much special attention Benny was getting as a result of his "obnoxious" behavior. At this point they were willing and eager to learn a more effective way of dealing with the problem.

The parents were asked to do a functional analysis for several reasons. First, since they indicated that the problem did not occur with a high frequency, we felt that this procedure would not be too taxing. When the problem behavior occurs with a high frequency, it is difficult to get the mediators to perform the more time-consuming functional analysis. Second, since both parents were unaware of the role that they might be playing in maintaining the behavior, the record-keeping procedure provided them with important feedback as to the effects they might be having on the problem. Here is a specific example of the interrelationship of behavioral assessment and treatment.

In most cases, a functional analysis is not the major source of mediator-collected data. Frequently the mediators are instructed to use the "three-term contingency" for a few days and then a shift is made to the easier, less time-consuming static analysis. A variety of static measurement techniques suitable for parents and teachers have been spelled out by Hall (1971). Those found most useful are the following:

1. *Direct measurement of a permanent product.* Many academic and nonacademic behaviors result in products which last long enough to be measured and recorded. Academic behaviors could include the percentage of items completed by a student or an entire class, the number of words correct on a spelling test, or the number of blocks stacked one on top of each other. Nonacademic behaviors could include the number of articles of clothing left around the house, whether or not toys have been put away, or whether or not the dishes were washed.

2. *Event recording.* Using this procedure, the observer simply counts the number of discrete occurrences of the behavior within a given time period. A teacher could count the number of times a child calls out during independent seat work; a parent could count the number of times one child hits another. Situations arise in which the time that the mediator is involved in event recording varies from day to day. A parent may be able to collect data for one hour on day #1 and five hours on day #2. If this cannot be avoided, the data can be compared from day to day by converting to a rate per minute (rpm) measure. The formula used for this conversion is:

$$\text{rpm} = \frac{\text{number of behaviors counted}}{\text{total number of minutes observed}}$$

Thus, if a mother observed her child from 6 P.M. to 8:30 P.M. and recorded 15 temper tantrums, the rpm measure would be rpm = 15 tantrums/150 minutes = .10.

3. *Time sampling.* If the problem involves some behavior which is ongoing and continuous, such as cooperative play, then time sampling is most appropriate. The specified period of observation is divided into equal intervals. The observer then notes whether or not the behavior of concern was occurring at the end of the interval (a case illustration of interval recording, in which the behavior is observed *throughout* the interval, is presented below). For example, if a teacher was interested in assessing a child's level of on-task behavior during a given 30-minute period, she could divide this time into equal intervals of 5 minutes. At the end of each 5-minute interval, she would look at the child and note whether or not he was on-task. When data are collected using the time-sampling procedure, the results are presented in terms of "percent of the time."

$$\text{Percent of the time} = \frac{\text{no. of intervals in which behavior occurred}}{\text{total number of intervals}}$$

For example, if a teacher used a 5-minute time-sampling procedure for 30 minutes and found a child to be on-task for two of the intervals, the results would be: Percent of the time on-task = 2/6 = 33%.

4. *Duration.* This measure is used to assess the length of time of a given problem. A teacher might be interested in the number of minutes it takes the class to settle down after a change in activities. Parents have often been concerned about the duration of their children's temper tantrums.

In order to maximize the cooperation of the parent or teacher, the therapist should outline the choices of data collection procedures available to them. The mediators are encouraged to choose a method with which they feel comfortable. Many failures occur at this point because the therapist "forces" a procedure on the mediator. Often after one week, they realize that another method is preferable and changes are made accordingly. In addition to providing the mediator with a choice, the therapist should also provide as much structure as is necessary. This may involve helping the mediator to design the data sheet or even having them printed up ahead of time. Data collection is also facilitated by role playing a problematic behavior so that the mediator can practice record-keeping. Finally, expectations are clearly specified by having blank graph paper available and indicating that the beginning of the next session will be spent in putting actual data onto the graph paper. If therapists make use of choice, structure, rehearsal, and clear expectations, parents and teachers are more likely to collect the sort of high-quality data that are basic to behavioral assessment.

Observations by the Therapist

Several investigators have made significant research contributions to behavioral technology and to the evaluation of behavioral treatment in the home and school (e.g., Eyberg & Johnson, 1974; Patterson, 1971). One of the unique characteristics of these large-scale research projects is their use of paraprofessionals (e.g., housewives and undergraduate students) to collect data in the child's natural environment or in laboratory analogues. Although this methodology has significantly advanced our scientific understanding of human behavior, it has unintentionally discouraged many therapists from venturing out of their office into the "real world." Because of the complex coding systems (a system may contain 30 or more codes), the amount of time to train observers, and the necessity for computer scoring, the task has been considered too formidable for use by those engaged in the delivery of services.

In our clinical work we have found observing the child in the context of the home and school to be invaluable during the initial stages of treatment. In order to implement naturalistic observations in the home and school, the methodology developed by behavioral researchers has to be greatly modified. Due to the high cost in terms of time and effort, we typically schedule one home and one classroom observation after the initial interview. The therapist should discuss the observations with those involved. The rationale, procedure, and ground rules should be clearly specified. It often helps to acknowledge that those being observed often feel awkward but that people tend to adapt to the observer's presence.

Home Observations

The best time to conduct a home observation is usually just before, during, and after dinner unless, of course, the problem under question is specific to a different time period. Dinner is a time when the family is together and interaction is at a maximum.

When the observer goes to the home at dinnertime, parents frequently assume the role of host and hostess and offer to set another place for dinner. This should be graciously refused with the explanation that it would be very difficult to work and be a dinner guest simultaneously. Instead, the parents are encouraged to ignore the observer and not discuss the observation until it is completed. The parents are instructed to discuss the observation with the children and simply explain that someone who is studying families is going to observe them. Since the interview with the parents has provided initial information about possible problems, the observer may want to do a functional analysis whenever that behavior occurs. In addition, complex sequences of behavioral interactions are sometimes revealed that may have been overlooked during the interview. Static analysis in terms of event recording, time sampling, and duration recording regarding the child's problem behaviors is also performed. These procedures are often applied to the parents and siblings as well.

A case illustration of a home observation with a family consisting of a mother, father, 7-year-old girl, and 9-year-old boy considered to be the problem child, follows. The initial interview proved to be less than informative, aside from suggesting that the parents were disorganized and knew little about child management, and that the household rules were applied inconsistently. Subsequent to the interview, a home visit was scheduled. After 10 minutes of observing the family and ignoring the children's demand for attention, the therapist began formal record-keeping.

Observations were made for 20 minutes using a one-minute interval recording procedure on each child. In interval recording, the behavior in question is observed for the entire length of the interval. This is to be contrasted with time sampling, where the behavior in question is only observed at the end of the interval. One child was observed for the first minute, the other child was observed for the second minute, and so on, until each child was observed for the 20 minutes. The behavior to be observed was labeled either "appropriate" or "inappropriate." Inappropriate behavior consisted of yelling, whining, and hitting. The interview with the parents enabled the parents and the therapist to define the problem adequately in this manner. The observer also collected data on positive and negative responses that the parents made toward the children. The actual data sheet used with this family is seen in Figure 3.1.

Above and below the recording for each child, data are entered

Figure 3.1 Sample Data Sheet of a Home Observation

Target Child:	Mike	Observer:	SBG
Other Children:	Jenny	Length of Interval:	one minute
		Sheet #:	1
		Date:	6-20-75

Setting:
Family having dinner, watching T.V. at same time. After dinner father and two children sit in family room while mother cleans up.

Behavior Codes:

I = inappropriate behavior
(hitting, whining, yelling)

A = appropriate behavior

Parent responses
− = disapproval
+ = approval
0 = nonevaluative comment

mother	−	−	−	−		−		−	
Mike	A	I	I	I	I	I	I	A	I	A
father				0						
mother		+		−						
Jenny	I	I	A	A	A	I	I	A	A	I
father										
mother	−	−	−			−	−	−	
Mike	I	I	I	I	I	I	A	I	I	I
father										
mother		+	+							
Jenny	I	I	I	A	I	A	A	I	I	I
father				0						

Comments:
Father seems to take nonverbal cues from mother to respond. His responses appear to be ineffectual.

regarding the responses made by the mother and father. The data revealed that the boy was behaving inappropriately 80% of the time and that his sister was behaving inappropriately 60% of the time. The high rate of deviant behavior for their daughter was a surprising finding to the parents. In addition, although both mother and father were present for the same time period, the mother made 15 responses to the children compared to her husband's 3. Of the mother's 15 responses, 11 were directed toward her son and only 4 toward her daughter. Of the 11 directed toward her son, all 11 were disapproving. Of the 4 toward her daughter, 3 were approving. The father's three comments (one directed toward his son, one toward his daughter, and one toward both of them) were viewed as nonevaluative. When these data were reviewed with both parents, they immediately began to realize more clearly the nature of the problem as well as the issues that needed to be dealt with in treatment.

When the observation is complete, the observer should inquire as to how typical their behavior was under this condition. Usually, the parents will report that, in the beginning, they held back but that, in general, there appears to be little distortion. Shortly thereafter, the results are analyzed and fed back to the parents. This sharing of results proves to be pivotal with regard to crystallizing the problems.

Classroom Observations

Considering the fact that visitors to classrooms are fairly commonplace, it is reasonable to assume that there is far less reactivity in this type of direct observation than in the homes. In practice, there is little substantive procedural difference between observations conducted in these two settings. The data sheet used for home observations can be used as is or modified to fit any special circumstances.

After the data have been analyzed, the results are shared with the classroom teacher and other appropriate school personnel. We suggest that the mediators be given direct feedback with regard to the problem child's behavior, their own behavior, and that of other children in the classroom. The importance of this became apparent to us during the early stages of development of this assessment procedure. Initially, the deviant child's peers had not been included in our naturalistic observations because we had simply not considered it to be important for clinical service. Following a classroom observation of a child referred for treatment, the teacher was told that the referred child was on-task 60% of the time. The teacher, totally unimpressed by our scientific data-collection procedure, seemed puzzled and asked: "Is that good or bad?" At that point, it became painfully obvious that we had overlooked a concept basic to developmental psychology: normative data. Nelson and Bowles (1975) state that high-quality normative data with regard to classroom behavior are sorely needed. Normative data not only aid in the identification and selection of deviant children, but also enable teachers and therapists to set realistic goals for intervention (Walker & Hops, 1976).

We have recently completed a study of naturalistic observations of four deviant and four normal children in two fourth-grade classrooms (Gordon & Keefe, 1977). The deviant children engaged in appropriate classroom behavior at a rate of 40%, whereas the normal children engaged in appropriate classroom behavior at a rate of 80%. Although there have been some attempts to collect such normative data in the school (Forness & Esveldt, 1974, Walker & Hops, 1976) and even in the home (Johnson, Wahl, Martin, & Johanson, 1973), more research is needed. Considering our present state of knowledge, those engaged in clinical service would do well to collect some type of normative data routinely for each class with which they are involved. Behavior therapists would be doing a more thorough assessment if, in addition to observing the deviant child, they would sample the behavior of several nonproblematic peers. These data would enable the teacher and therapist to assess the severity and nature of the problem as well as to establish realistic goals for treatment.

Matching Treatment to Client

Assessment of Environmental Control

One of the guiding principles of behavior therapy is that human behavior is largely a function of the environment: treatment must focus on those environmental contingencies which are discriminative and/or reinforcing stimuli for the target behavior. This concept is important to successfully match the treatment plan to the client. It is common to encounter parents whose children behave in a deviant fashion in school but who exhibit little deviant behavior in the home. Behavioral treatment then suggests that the initial focus of intervention be in the school and not in the home.

The situation is further complicated if the school does not perceive a problem or if it prefers not to seek the aid of a behavioral consultant. An assessment of these controlling factors should enable the therapist to design a more suitable treatment plan.

Interpersonal Problems Between Parents

Since the success of behavioral treatment with children is maximized by the degree to which the child's parents are willing and able to work together, the quality of the parental relationship may set some constraints on the setting of goals and the strategies used to meet these goals. Miller (1975) states:

> There are many reasons for the parents' failure to work together in a supportive way. Very frequently, interparental inconsistency reflects problems in the marital relationship. The parents' lack of agreement or cooperation with each other in parent training may indicate the presence of more general negative or ambivalent attitudes about each other which interfere with their response to treatment. Such parents may show intense conflicts about any joint decision involving the family life. . . . (p. 98)

The use of a brief screening instrument such as the Locke-Wallace Marital Adjustment Scale can alert the therapist to potential problems in this area. Often, this information is not seen until treatment is underway. This is where the assessment of ongoing therapy allows the therapist to reformulate treatment strategies and thereby more effectively match treatment to client.

Intrapersonal Problems of Mediators

The mediators in the child's environment are subject to all the same disturbances and upsets of any human being. The intrapersonal problems of mediators may interfere with the direct application of social learning principles to modify the child's behavior. Inadequate responsiveness is one of the most difficult interference factors to identify and resolve (Miller, 1975). This is characterized by the mediator's contradictory message: "Help me but don't expect me to change." Typically this information is not picked up at the initial session but rather, after the parent or teacher realizes that he must change his behavior prior to a change in the child's behavior. At this point, treatment needs to take a different course to more effectively deal with this problem. Other intrapersonal problems of mediators, such as depression or severe anxiety, are in need of assessment so that treatment can be personalized. This point was dramatically demonstrated recently when two of the authors were seeing two families, each with an elective mute child. Although the presenting problems were identical, i.e., the child spoke at home but not at school, the family contexts were totally dissimilar. In one case the parents' marital relationship was excellent with no evidence of any intrapersonal problems. The second case, however, presented a picture of severe marital discord and a mother suffering from severe anxiety. In spite of the identical problem of the child's not speaking in school, the cases required very different treatment strategies.

Resources of the Child

Very often behavioral assessment has overlooked the child's own resources. Recently, self-control with children has been receiving greater attention by behavior therapists (Graziano, 1975). With the development of this approach, we need to consider the child's behavioral assets, the child's motivation to become active in the process of treatment, and the child's biological and sociological factors which may facilitate or impede change. The case of a 12-year-old boy diagnosed as having minimal brain dysfunction illustrates this point. This child was referred, not because of any academic problems, but because of his poor peer relations. Specifically, he was continuously teased and taunted by a number of his peers. His response was to become anxious, which served to exacerbate some of his fine-motor difficulties, which resulted in even more teasing. The child was upset by this state of affairs and was highly motivated to become active in the treatment process. He was given some simple instruction in social learning concepts and realized how he used to intermittently reinforce the teasing by returning insults and acting silly. With this realization, he was able to deal successfully with the problem through self-monitoring, graphing, ignoring, and emotive imagery (Lazarus & Abramovitz, 1962).

Assessment of Ongoing Therapy and Evaluation of Therapy

Continuous Data Collection by Mediators

A major strength of behavioral assessment and therapy is that assessment does not stop when therapy begins. Rather, assessment is an ongoing, continuous process. The mediators continue to collect relevant data throughout. Their data become the focal point for making decisions regarding treatment. When parents and teachers report satisfaction with the program, they can be instructed to collect data on a less frequent basis.

Periodic Assessment by Therapist

The therapist may want to assess the present state of the problem through means other than mediator-collected data. The therapist may find it useful to periodically perform an observation of the child in the home, school, or laboratory. In addition, periodic readministration of several of the paper-and-pencil measures may provide additional information. These overall results may either confirm or contradict the results from other sources. Although multiple sources of data collection are viewed as necessary in the behavioral assessment of children (Eyberg & Johnson, 1974), this approach is not without its problems (Gordon, Lerner, & Keefe, in press; Johnson & Eyberg, 1975). We hope to find consistency across all of our multiple measures, but unfortunately the data are often unkind and do not make it easy for behaviorally oriented clinicians. Behavior often varies from setting to setting and data from varying sources of information are less than consistent (Johnson & Christensen, in press). These conundrums could almost force a behavior therapist to an analyst's couch! We are in need of research that looks at the relationship between these various measures. Until that time, those engaged in the practice of behavior therapy should be aware that a precise science of behavioral assessment with children does not yet exist.

Generalization and Maintenance

The issues of generalization across settings and behaviors as well as maintenance of behavior change have been thoroughly reviewed by Conway and Bucher (1976). The outcome questions that need to be considered are: What behavior was changed? Under what stimulus conditions was it changed? For how long was it changed? The information needed to answer these questions becomes available when the therapist makes use of the procedures discussed in this chapter. Finally, a closing conference with the multiple mediators involved in the treatment allows for closure and an examination of steps to follow, should problems occur in the future.

CHAPTER 4

Behavioral Assessment with Adult Outpatients

In Chapter 4, we examine practical methods for the behavioral assessment of adult outpatients and we review critically methods that can be applied to a wide assortment of presenting problems. These methods include: (1) the interview; (2) questionnaires; (3) observation; and (4) psychophysiological analysis.

The Interview

The interview is probably the most widely used method of assessing adult outpatients. Virtually all contact with adults occurs in interview sessions. The interview serves as a setting for: (1) establishing a relationship with the client; (2) gathering information; and (3) influencing and directing behavior change efforts. Thus, the interview holds a pivotal position in the process of behavioral assessment.

Establishing a Therapeutic Relationship

The importance of establishing a relationship with the client has been recognized by behaviorally oriented clinicians for many years (Kanfer & Phillips, 1970). The therapist consistently structures the interview to facilitate the development of this relationship. Early in therapy, it is best to use a low level of interviewer activity (Sullivan, 1954). Repetition of the client's last few words in a sentence, paraphrasing, and clarification are interviewer responses that help a therapeutic relationship develop quickly and make it easier for the patient to describe his problems in his own terms. As the patient becomes more comfortable and rapport is established, more active interviewer responses are appropriate.

In an effort to be "objective," novice behavior therapists often act cold and aloof in clinical interview sessions. This is a mistake. Research indicates that the quality of the therapeutic relationship can directly effect the outcome of behavioral treatment approaches, such as systematic desensitization (Morris & Suckerman, 1974). The relationship between therapist and patient is viewed as important in the process of behavioral assessment. In the interview, appropriate levels of therapist warmth and understanding are necessary for this relationship to develop.

Gathering Information

Interviews are also used to gather information about the client. During the first few interviews, background data are obtained. These early interviews also provide a direct sample of the client's behavior. Many of the behaviors sampled in the traditional mental status exam (Martin, 1972) are relevant. For example, the client's manner of speech, use of gestures, eye contact, and physical appearance provide clues to the behavioral clinician. These clues suggest problems that initially may not have been reported by the client himself. For example, the client who reports he is afraid of social situations may demonstrate in the interview that he lacks the necessary assertive and social skills. This information may very well direct the course of subsequent therapy towards alleviation of this deficit.

In the process of assessment, clients are often instructed to keep records. When working with adult outpatients, these records are usually reviewed in interview sessions. For example, responses to questionnaire items can be reviewed with attention directed towards possible omissions or misunderstandings. Data gathered through self-observation can be plotted on simple graphs and then discussed.

In the interview, the therapist gathers information. Because he does this openly and is willing to offer complete explanations of rationale and techniques, he serves as a model for the client. The value of systematic data collection becomes apparent over the course of therapy. Thus, in gathering data the behavior therapist teaches the client a systematic approach that can be applied to deal with a variety of problems in living.

Influencing and Directing Behavior Change Efforts

In the interview setting, the behavior therapist systematically uses his own influence as a social reinforcer to influence and direct behavior change. Early in assessment, the therapist may refrain from the use of such reinforcement to minimize bias (Thomas, 1973). However, as assessment progresses and goals and behavioral targets become clearer, the therapist makes frequent use of reinforcement principles in the interview. Smiles, nods, and verbal praise are used to reward progress in the therapeutic direction. In addition, extinction is also used in interview sessions to decrease the frequency of inappropriate behaviors (for example, complaining or obsessive ruminating). When applying an extinction procedure, the therapist ignores or fails to respond to the patient's behavior. The judicious combination of reinforcement and extinction procedures in the interview can be a very effective clinical tool. Consider the following transcript taken from an interview with a phobic patient:

> THERAPIST: Let's begin today by reviewing your progress on the homework assignment.
>
> PATIENT: Well, I completed all ten items that we picked from the hierarchy—
>
> THERAPIST (*interrupting to reinforce this*): That's good.
>
> PATIENT: Yes, but I really had a difficult time on the last item when I rode that elevator. I don't know, I don't think I'll ever overcome these fears.
>
> (*Since the therapist has been made well aware of the patient's feelings of hopelessness in earlier sessions, he decides that a focus on this topic is unproductive. He therefore chooses to use extinction and ignore the client's negative statements. He tries instead to encourage the patient to describe his actual behavior in the situation.*)
>
> THERAPIST: Tell me what actually took place when you worked on that item. How did you behave or act?
>
> PATIENT: There's not much to tell. I just feel lousy about it.
>
> THERAPIST (*uses extinction again, ignores this last comment and returns to his last point*): Take some time to think it over. What were you doing before, during, and after that elevator trip? It might help to review your records in your notebook.
>
> PATIENT (*pause*): Well, first of all I got into the elevator, then I rode to the fourth floor and got off. I felt anxious and upset during the ride up. It was the toughest task in the assignment.

Behavioral Assessment with Adult Outpatients

(*This patient is obviously not in the habit of reinforcing himself for behavior change. He has chosen to start the interview off by focusing on major failure rather than on his successes. A good idea would be to try to change this pattern in the interview session itself.*)

THERAPIST: Tell me, what did you like about the way you handled this task?

PATIENT (*laughing*): Not much of anything (*pause*). The only thing that I can think of was the fact that I stuck with it until I got all the way to the fourth floor.

THERAPIST (*reinforces this statement*): Very good. You must be pleased about doing that.

PATIENT: Well, yes I guess I am. Four months ago I wouldn't have even gotten near that elevator.

THERAPIST (*probing for another appropriate behavior the patient could be reinforced for*): Was there anything else that you liked about the way that you handled this part of the assignment?

PATIENT: Well, the only other thing is something I did before the elevator trip.

THERAPIST: What was that?

PATIENT: I used the relaxation procedure in my office.

THERAPIST (*reinforces this*): That's good. How did that help you? Try to be as specific as you can.

(*Talking at length about an appropriate behavior tends to act as a reinforcement.*)

PATIENT: For one thing, I was able to relax the muscles in my shoulders and neck, so that I didn't get that headache that always bothers me in these situations.

THERAPIST: That must have been a relief!

PATIENT: It was. I also used the relaxation in dealing with one of the easier tasks—climbing the stairs in my apartment to the sixth floor.

THERAPIST: Good (*reinforces*). Let's talk about how you handled that task.

In this case, the therapist was able to direct the patient towards recognition of his behavioral assets. Oftentimes, patients tend to overlook appropriate behavior to focus on their problems and failures. Discussion about these inappropriate behaviors is likely to be upsetting and counterproductive. In behavioral therapy with adult outpatients, environmental control is difficult to achieve. Working directly with the patient in his natural environment is also usually too costly and intrusive. The client must, therefore, assume responsibility for completing homework assignments and carrying out interventions. The therapist's use of reinforcement principles in the interview is one of the most important factors that influence and guide the client towards completion of such tasks.

Issues in the Use of the Interview

The interview has many advantages. It is one of the most practical and economical assessment procedures available. In behavioral assessment, the interview can be used for a number of purposes. These facts undoubtedly contribute to the popularity of the interview in work with adult outpatients.

The validity of interview data, however, is often suspect. The clinician needs to be cautious in accepting verbal reports at face value. A patient may report a change in behavior when, in reality, no change has taken place. To overcome this problem, data gathered in interviews with the patient should be checked for accuracy against data from other sources, such as the patient's spouse and peers or laboratory observations of the patient. Periodic checks help ensure that the interview data and the decisions based on them are both meaningful and valid.

Questionnaires

For years, behavioral clinicians avoided the use of questionnaires and relied mainly on data gathered through observations. As behavioral approaches are used more and more extensively with adult outpatients, the utility of questionnaires is becoming more and more apparent. At present, a variety of questionnaires are typically used in work with adults, including questionnaires designed to gather information on: (1) behavioral antecedents; (2) response characteristics; (3) response consequences; and (4) general history.

Behavioral Antecedents

As pointed out in Chapter 1, a major tenet of behavior therapy is that behavior is situationally specific. Environmental settings are important sources of control over behavior. In behavioral assessment, questionnaires can help identify specific environmental antecedents which clients view as significant.

The Fear Survey Schedule (FSS) is a good example. The FSS asks the client to rate the severity of his fear reaction to a variety of stimuli. Items included in the FSS are many of the stimuli that produce anxiety in outpatients, e.g., airplanes, being in an elevator, and journeys by car. Several forms of the FSS have been developed (Hersen, 1973). For purposes of discussion, we shall consider only the FSS-III (Wolpe & Lang, 1964).

The FSS-III is a 72-item schedule developed for use in clinical practice. The FSS-III is scored in several ways. The clinician may examine ratings for one or two items of central importance. When a specific fear is the referral problem, use of a single item on the FSS is adequate for assessment purposes (Lang, 1969). A second approach is to compute a total score. Total scores are obtained by summing ratings for each item. Ratings are on a 5-point scale (1 = not at all; 5 = very much). Thus, total scores range from 72 to 360. The higher the score, the more pervasive the fear response.

Successful therapy should result in a decrease in the total number of fear-producing stimuli sampled on the FSS.

Response Characteristics

Questionnaires are also used to collect data on the patient's perceptions of his own behavior.

Rating Scales

Clients may be asked to make ratings of their own behavior. Ratings may be taken on a variety of dimensions, the poles of which can be determined by discussion with the client (Gottman & Lieblum, 1973; McCullough & Montgomery, 1972). One of the most popular rating scales is the subjective discomfort scale, or SUDS (Wolpe & Lazarus, 1966; Wolpe, 1969). The SUDS scale runs from 0 to 100. The zero point on this scale denotes a state in which the client feels as relaxed and free from anxiety as possible. The top or 100 point on the scale refers to a state in which the client feels as anxious and tense as possible. Ratings on the SUDS are easily taken in a variety of situations. The client may rate his degree of discomfort while actually engaging in anxiety-provoking behavior—for example, while standing in an elevator located in a tall building. Clients may also rate the discomfort they experience while imagining themselves engaging in certain behaviors. Fryrear and Weiner (1970), for example, asked a nursing student who had a strong fear of dissecting live animals to rate her feelings in

SUDS as she imagined herself performing physiological demonstrations, such as cutting off the head of a live frog.

Although we have only discussed rating scales used by clients, it is often helpful to have the therapist or an independent observer rate the client's behavior along identical dimensions.

Checklists, Inventories, and Surveys

Below is a list of some of the common checklists, inventories, and surveys used by behavior therapists. The questionnaires listed differ in form and psychometric sophistication. They are used in assessment as an extension of the interview (through examination of answers to particular items) or to compare the client's responses to those made by other individuals of the same age and sex (through examination of standardized or factor scores). Research indicates that many of these questionnaires are both reliable and valid when used with adult outpatients. We recommend that the reader familiarize himself with the relevant research before using any of these questionnaires on a routine basis.

Questionnaires Used in Assessment of Behavior

Problem Area: DEPRESSION
Beck Depression Inventory (Beck, 1967)
D Scale MMPI
Depression Adjective Checklist (Lubin, 1965)
Self-Rating Scale (Zung, 1965)

Problem Area: ANXIETY
Affect Adjective Checklist (Zuckerman, 1960)
S-R Inventory of Anxiety (Endler, Hunt, & Rosenstein, 1962)
Taylor Manifest Anxiety Scale (Taylor, 1953)

Problem Area: TEST ANXIETY
Test Anxiety Questionnaire (Mandler & Sarason, 1952)
Test Anxiety Scale (Sarason, 1958)
Achievement Anxiety Test (Alpert & Haber, 1960)
Suinn Test Anxiety Behavior Scale (Suinn, 1969)

Problem Area: ASSERTIVENESS
Wolpe-Lazarus Assertiveness Questionnaire (Wolpe & Lazarus, 1966)
Rathus Assertiveness Scale (Rathus, 1973)
College Self-Expression Scale (Galassi, DeLo, Galassi, & Bastien, 1974)

Problem Area: SOCIAL SKILLS
Social Avoidance and Distress Scale (Watson & Friend, 1969)
Social Activity Questionnaire (Arkowitz et al., 1975)

Problem Area: OBSESSIONS AND COMPULSIONS
Leyton Inventory (Cooper, 1970)

Problem Area: OBESITY
Eating Pattern Questionnaire (Wollersheim, 1970)
Physical Activity Scale (Schifferes, 1966)

Problem Area: MENSTRUAL PAIN
Menstrual Activity Scale (Tasto & Chesney, 1974)
Menstrual Symptom Questionnaire (Chesney & Tasto, 1975)

Response Consequences

Relatively few questionnaires have been developed for the purpose of identifying the natural consequences of problematic behavior. Most often, information regarding behavioral consequences is identified through interview methods, such as reviewing the client's typical day, or by direct observation. Two questionnaires have been developed, however, that help identify stimuli that are reinforcing to the client and can be used in treatment programs to make behaviors more likely to occur: the Reinforcement Survey Schedule and the Pleasant Events Schedule.

The Reinforcement Survey Schedule

The Reinforcement Survey Schedule (RSS) was developed by Cautela and Kastenbaum (1967). This questionnaire is divided into four sections that cover a broad range of potentially reinforcing stimuli. Section I asks for ratings of the degree of pleasure that a number of tangible items and situations bring, e.g., eating ice cream and drinking beer. Section II explores the reinforcement experienced by the respondent in a variety of activities, e.g., shopping and being praised. Section III focuses on social situations which may arise. Section IV asks the respondent to list things that he does five, ten, fifteen, and twenty times a day. Section IV is based on the Premack Principle (Premack, 1959), which states that behaviors of high probability (e.g., talking with spouse) reinforce behaviors of low probability (e.g., approaching a feared situation).

Several case studies reported by Cautela and Wisocki (1969)

illustrate the use of the RSS in selecting reinforcers for treatment. One client, who indicated a high preference for classical music on the RSS, overcame his fear of urination more quickly when given records as reinforcement than when given only in vivo desensitization. A second client, who reported a strong preference for monetary reinforcement on the RSS, rapidly overcame his driving phobia when a dollar was given to him for each mile he drove away from home. Similar studies have indicated that the RSS can aid in the assessment of covert reinforcers, such as imagining a pleasant activity or scene (Cautela & Wisocki, 1969; Cautela, 1970; Epstein & Peterson, 1973).

The Pleasant Events Schedule

The Pleasant Events Schedule (PES), described by MacPhillamy and Lewinsohn (1971), has also been used to identify specific events and activities that are reinforcing. Lewinsohn (1975) has used the PES extensively in the behavioral management of depressed patients. The PES helps point out to depressed patients the fact that they are not engaging in many activities that they consider pleasant. Treatment is then aimed at having patients systematically increase the number and kind of reinforcing activities they engage in.

General History

History questionnaires are a convenient method of gathering background information. Usually, they are administered early in therapy. A good example of a general history questionnaire is Lazarus' Multimodal Life History Questionnaire, published for the first time as Appendix B of this book. It covers a wide range of topics, including developmental history, occupational data, sexual history, marital history, family background, and current life problems. Several questions focus on specific behavior patterns. Clients are asked, for example, to estimate the current severity of their problems using a 7-point rating scale (Question 2C).

Two assumptions underly the use of such general history questionnaires (Kanfer, 1975):

> 1. Sampling a wide range of content areas will provide cues for the location of problems.
> 2. The patient's behavior (style) in handling various aspects of his life will reveal the generalized patterns that he uses to solve interpersonal and intrapersonal problems. (p. 78)

A second type of history questionnaire is the problem history questionnaire, developed to gather extensive information on one presenting problem. A variety of forms are available. Examples include the Dating History Questionnaire (Curran & Gilbert, 1975); the Smoking History Questionnaire (Keutzer, 1968); and the Stuttering History Questionnaire (Martyn & Sheehan, 1968). Problem history questionnaires ask the client to indicate situational determinants that appear to influence his behavior. Ratings of severity are also obtained. The clinician may wish to employ individual items from such questionnaires repeatedly over the course of therapy.

Issues in the Use of Questionnaires

The major advantage of using questionnaires with adult outpatients is that they are an inexpensive way of gathering a great deal of information. The questionnaires reviewed are easy to administer and score. They tend to ask direct questions about highly specific activities, events, and behaviors. The use of such direct and simple measures has a basis of support in the research literature (Mischel, 1968).

There are several practical considerations in the use of questionnaires during behavioral assessment. Because questionnaires are easy to administer and score, the clinician may be tempted to combine many of them in a battery and administer them routinely to all patients. This is not recommended. The administration of too many tests in combination with other data-gathering procedures (e.g., self-observation and naturalistic observation) overwhelms patients, who may lose sight of the fact that the results of each questionnaire are to be used in planning treatment. Lewinsohn (1975) suggests that it is best to administer one test instrument at a time and to make the next appointment contingent on completion of that test. He also recommends that the therapist only administer those questionnaires that promise to directly contribute to formulating treatment plans.

Questionnaire reports should be accepted with some degree of reservation. The validity of questionnaire data, like that of data gathered through interviews, is questionable. The relationship between responses on a questionnaire and actual behavior may be poor. Azrin, Holz, and Goldiamond (1961) have argued that direct and objective measures of behavior are always needed to determine the validity of questionnaire responses.

To summarize, questionnaires provide the clinician with an

expedient aid in behavioral assessment. There are, however, major difficulties associated with the uncritical use of self-report data gathered by means of questionnaires. The clinician would do best to use questionnaires as one component of a more comprehensive behavioral assessment.

Observation

Observation is often considered the fundamental method of behavioral assessment (Lipinski & Nelson, 1974). In the assessment of adult outpatients, three forms of observation are used: self-observation, naturalistic observation, and laboratory observation.

Self-Observation

Self-observation is probably the most widely used assessment technique employed with adult outpatients. Self-observation involves three steps (Thoreson & Mahoney, 1974). First, the client learns to discriminate between the occurrence and nonoccurrence of a particular behavior, e.g., eating a calorie-rich food, smoking a cigarette, having an irrational urge. Second, the client records the behavior. Third, the client summarizes the data and compares it to some standard. A variety of recording strategies are employed in self-observation, including the daily notebook, counters, index cards, and charts.

The Daily Notebook

The daily notebook is a comprehensive record-keeping procedure. The notebook is especially helpful in dealing with clients who are unaware of the relationship between their own behavior and environmental antecedents and consequences. The client's task in keeping a daily notebook is fairly simple. He is asked to make an entry in the notebook each time the problematic behavior occurs. The entry should include a description of what happened before, during, and after the occurrence of the behavior.

Having clients keep a daily notebook has several advantages. It tends to familiarize them with the process of record-keeping. Clients frequently respond to the instructions by saying, "Oh, you want me to keep a diary." The task is a familiar one, and one that they may have tried in the past. Because of this, clients are more likely to accept and actually use the procedure. The daily notebook yields data that are rich in detail.

Counters

Counters are often used in self-observation. Golf counters that can be worn on the wrist or concealed in a convenient pocket are especially popular (Lindsley, 1968; Hannum, Thoreson, & Hubbard, 1974). Counters can also be fashioned from items found around the house. Thoreson and Mahoney (1974) have suggested the use of knitting tallies which fit over the tip of a pencil. Making tears in the corners of pieces of paper, switching marbles from one pocket to the other, or whittling notches in a dowel are just some examples of homemade alternatives to manufactured counters (Watson & Tharp, 1970). K. Mahoney (1974) has described an ingenious counter that can be easily fashioned and resembles handcrafted leather jewelry. This counter is unobtrusive and can be used to simultaneously monitor multiple behaviors such as lip biting, finger picking, and smoking (Epstein & Hersen, 1974).

Index Cards

One of the most popular materials for self-observation is the 3" × 5" index card. Index cards are easily carried in a pocket or pocketbook. The format for recording varies. An individual who complains of "writer's block" may be asked to record the total number of minutes spent writing daily. A phobic may use the card to make ratings of tension levels in everyday situations. A smoker can cut the card in half, slide it into the cellophane wrapper of a cigarette pack, and write down the time at which he smokes each cigarette.

Charts

Simple charts or data sheets are handy for self-recording. Clinicians either fashion these to meet the specific needs of a case or rely on standard charts for use with commonly encountered clinical problems. For example, Table 4.1 presents a copy of a chart used by Bradner and Surwit (personal communication, 1976) in the assessment of migraine headaches. During the early stages of assessment, patients fill in the top three rows of the chart. Headache intensity, presence of an aura, and medication are recorded on a daily basis. Later on, patients are provided with a temperature-sensitive strip for biofeedback training (available from Medical Device Corp., P.O. Box 217, Clayton, Indiana 46118) and are given written instructions on how to use the strip for recording during home biofeedback practice sessions. Temperature changes occurring during home practice are also recorded on the chart.

Table 4.1
A Chart for the Assessment of Migraine Headaches

Name		Monday	Tuesday	Wednesday	Thursday	Friday	Saturday	Sunday
HEADACHE INTENSITY								
0 = no headache	morning	___	___	___	___	___	___	___
1 = slight headache								
2 = moderate headache	afternoon	___	___	___	___	___	___	___
3 = fairly severe headache								
4 = very severe headache	evening	___	___	___	___	___	___	___
5 = most severe headache								
AURA								
0 = no signs that headache	morning	___	___	___	___	___	___	___
is coming								
1 = some sign that headache	afternoon	___	___	___	___	___	___	___
may be coming	evening	___	___	___	___	___	___	___

Name _____

	Monday	Tuesday	Wednesday	Thursday	Friday	Saturday	Sunday
MEDICATION TAKEN							
Type 1 ___name___ mg.							
Type 2 ___name___ mg.							
Type 3 ___name___ mg.							
Type 4 ___name___ mg.							
HOME PRACTICE SESSIONS							
Begin							
time							
number							
color							
End							
time							
number							
color							

Reprinted by permission of Marilyn Neyer-Bradner and Richard S. Surwit.

The main advantage of charts is that they simplify record-keeping. The clinician is encouraged to develop forms suited to the demands of his own practice.

A number of different characteristics of behavior can be monitored during self-observation. Counters are typically used to record the frequency of behavior. Index cards are often used to monitor the intensity or duration of behavior. The daily notebook is employed to assess two or more characteristics of behavior at once, such as frequency and duration.

Table 4.2 presents examples of problem behavior frequently encountered in working with adult outpatients and lists some of the typical response characteristics monitored during self-observation. The clinician may wish to consult the table and references provided in formulating self-observation strategies for a given patient.

Naturalistic Observation

Naturalistic observations are often used in the behavioral assessment of children (see Chapter 3). Because of practical problems, these observations are not as widely used with adult outpatients. In this section, we review interesting naturalistic observation strategies that can be adapted to the demands of clinical practice with adults.

Home Observations

Lewinsohn and his co-workers (Johansson, Lewinsohn, & Flippo, 1969; Lewinsohn, Alper, Johansson, Libet, Rosenberry, Shaffer, Sterin, Stewart, & Weinstein, 1968) use home observations extensively in the behavioral assessment of depression. The procedure for home observations is straightforward. Observers visit the client's home for an hour around mealtime. The visit is scheduled at a time when all family members can be present. The family is told to limit themselves to one area of the home, for example, the kitchen, dining area, and living room. Family members are asked to refrain from watching television and to ignore the presence of observers. Interactions occurring among family members are recorded immediately prior to, during, and after the meal.

A major question in conducting any naturalistic observation is "How are the behaviors to be coded?" For home observation, Lewinsohn uses a complex code (Lewinsohn et al., 1968) requiring trained observers (Libet & Lewinsohn, 1973). Practicing clinicians prefer simpler coding systems. These can be developed to fit the needs of a specific case and may include as few as two categories of behavior.

Table 4.2
Examples of Problem Behaviors and Typical Responses
Monitored in Self-Observation

Problem Area	Response Characteristic Monitored & Reference
Alcoholism	Number of alcoholic beverages consumed per day (Miller, 1972)
Lip biting	Number of bites per day (Ernst, 1973)
Functional diarrhea	Number of bowel movements per day (Furman, 1973)
Excessive urination	Number of urinary episodes per day (Yates & Poole, 1972)
Compulsive cleaning	Duration of time spent cleaning (Wisocki, 1970)
Compulsive checking	Duration (minutes) of checking (Melamed & Siegel, 1975)
Headache pain	Intensity of headache pain on 5-point rating scale (Budzynski, Stoyva, & Adler, 1970)
Epileptic seizures	Number of seizures per day (Johnson & Meyer, 1974)
Exercise	Number of aerobic points per day (Kau & Fischer, 1974; Cooper, 1970)
Insomnia	Number of hours spent sleeping (Borkovec, Kaloupek, & Sloma, 1975)
	Number of minutes to sleep onset; amount of medication used (Evans & Bond, 1969)
Obesity	Weight in pounds (Romanczyk, Tracey, Wilson, & Thorpe, 1973)
	Caloric level per day (Stuart & Davis, 1972)
Stuttering	Number of stuttering episodes per day (Azrin & Nunn, 1974)
Obsessive thoughts	Number of half-hours per day in which at least one obsessive thought occurred (Hackman & McLean, 1975)
Agoraphobia	Number of minutes spent by self away from home (Emmelkamp & Ultee, 1974)

Home observations have several advantages (Lewinsohn, 1975). First, patterns of interpersonal behavior are revealed which are critical to case formulation. A home observation may indicate that a client's only "topic" in conversation is his "mental illness," or that he interacts only minimally with his spouse. Second, the scheduling of home visits gives the client the clear message that his behavior is the main focus of assessment. The client is encouraged to become more aware of his own behavior. Third, home visits provide a convenient opportunity to meet significant others in the client's life, such as his spouse and children.

In Vivo Tests

In vivo tests are designed to assess phobic avoidance behavior in actual fear-producing situations. Clients reporting irrational fears of objects or situations are asked to expose themselves to feared stimuli. The strength of the phobic response is measured along dimensions such as: (1) physical proximity to the feared stimulus; (2) duration of time taken to approach the stimulus; and (3) number of items completed in a response hierarchy.

Behavioral clinicians find themselves conducting in vivo tests in a variety of places. Elevators, escalators, and motor vehicles are common sites for such tests. Leitenberg and Callahan (1973) used a fire escape to test acrophobics. The fire escape was three stories high and had open-grate iron stairs. Clients climbed as high as they could and still not feel undue anxiety; they stayed for as long as possible. The number of steps climbed was taken as a measure of the strength of the phobic response. In vivo tests of agoraphobic patients can be conducted from the therapist's office (Emmelkamp & Ultee, 1974; Everaerd, Rijkin, & Emmelkamp, 1973; Watson, Mullett, & Pillay, 1973). Patients are told to go out of the office into the street, walk as far away as they can, and stay until they feel unduly tense. The length of time taken or the distance traveled from the office provides an indication of the strength of the fear.

Graded response hierarchies help quantify the results of in vivo tests. Response hierarchies are particularly useful if the test is repeated at regular intervals throughout therapy. In response hierarchies, items related to a given fear are arranged in order of difficulty. The precise order is determined through interviews with the client. A hypothetical typical hierarchy for a driving phobic follows. In it, 1 = the least difficult task; 5 = a moderately difficult task; and 10 = the most difficult task.

Response Hierarchy of Driving Phobic

Order of Difficulty	Fear-Related Item
1.	Get into car, start engine
2.	Back car out of driveway
3.	Drive one city block
4.	Drive three miles of uncongested streets
5.	Drive three miles of busy streets
6.	Drive superhighway for five miles
7.	Drive superhighway for 20 miles
8.	Drive superhighway for 100 miles
9.	Drive superhighway for 200 miles
10.	Drive superhighway for full day

Figure 4.1 A Map Showing Hierarchy Items of a Patient with Fear of Driving Distances

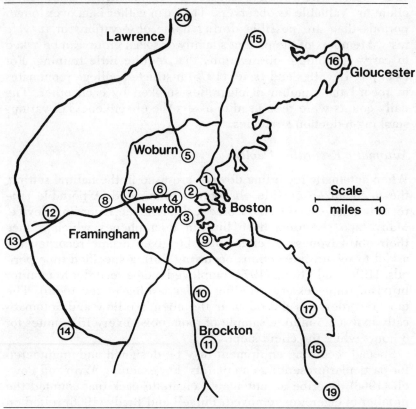

Hierarchies are scored by counting the number of items completed. With certain clinical problems, maps provide a helpful visual display of progress through a hierarchy. Figure 4.1 is a map used with a patient who had a profound fear of driving distances. The patient, who lived in Boston, circled each location she drove to in the course of completing a hierarchy. The number of the hierarchy item was written in the center of a circle. Inspection of the map shows that in the early stages of therapy, driving was restricted to the local Boston area (low numbers). As therapy progressed, the patient was able to drive to the suburbs and beyond (higher numbers).

The construction of hierarchies is a basic component of desensitization therapy. The reader may wish to consult references on systematic desensitization to familiarize himself with this aspect of behavioral assessment (Wolpe & Lazarus, 1966; Wolpe, 1969).

Observations by Family or Peers

Individuals who spend a considerable amount of time with a client are valuable as observers. They can gather data over longer periods than are possible during home observations or in vivo tests. Members of the patient's family or social group can be asked to carry out simple observations that require little training. For example, Lewittes and Israel (1975) instructed college roommates to record the number of cigarettes smoked by each other. The daily counts were then used to assess the effectiveness of various smoking-reduction strategies.

Automatic Recording Devices

When automatic recording devices are used in the natural setting, the need for an outside observer is eliminated. A portable tape recorder, for example, is a practical automatic recording device. Many tape recorders have built-in microphones which reduce their obtrusiveness. A client may be given a tape recorder and asked to record interactions occurring during specified time periods. Hiller and Strang (1973) employed a tape recorder to monitor bruxism (unnecessary grinding and gnashing of the teeth). The tape recorder was placed near the client's pillow and automatically took a 1.5-minute sample of room noise every 15 minutes for 6 hours while the client slept.

Special recording equipment may be designed and engineered for particular presenting complaints. For example, Azrin and Powell (1968) describe an automated cigarette pack that counted the number of cigarettes removed. Purcell and Brady (1966) relied on miniature radio transmitters to monitor the verbal behavior of adolescent patients. The development of nondetectible audio and visual devices, or "bugs," increases the possibilities for applying recording devices in the natural environment. Ethical problems may be raised in the application of these techniques. While the client has given his *informed consent* for recording, those with whom he interacts may not have done so. Automatic recording devices should never be used without the full knowledge and consent of all parties involved.

Naturalistic observations have several advantages. The detail and immediacy they provide helps pinpoint important parameters of behavior. Information that the client may have failed to report is available. Settings impose some astonishing constraints on human behavior. Naturalistic observations indicate clearly the situational factors controlling behavior. The patient who drives at reckless speeds on busy interstate highways may find crossing a suspension bridge so anxiety-provoking that he slows down to a crawl. Directly observing such phenomena restricts the tendency on the part of the clinician to overgeneralize from other sources of data, such as interviews or questionnaires.

Laboratory Observation

Direct measures of behavior also can be obtained by observing the client in the laboratory. The term "laboratory" refers to a room where observations can be made in a controlled manner. Laboratories used in behavioral assessment are either simply equipped with a few pieces of furniture, a one-way mirror, and an intercom, or are more elaborately outfitted with film cameras, audiotaping or videotaping equipment, and special electromechanical devices.

Minimal instructions to the client often are all that is needed to have him perform his problem behavior in the laboratory. Stutterers may be asked to read a passage from a book (Gray, 1965). Two people who are anxious about dating may be told to carry on a conversation to "get to know one another" (Glasgow & Arkowitz, 1975). The laboratory environment can also be structured to elicit specific behaviors of interest. Role playing and simulated environments can be used for this purpose.

Role Playing

In some laboratory observations, the client is asked to pretend he is in a particular situation and to role play how he would respond. In the Behavioral Assertiveness Test (Eisler, Miller, & Hersen, 1973), for example, an assistant administers the test by playing various roles and prompting responses by the client. Interpersonal situations of several kinds are sampled, for example:

> Scene 1
>
> *Narrator:* You're in a crowded grocery store and in a hurry. You pick one small item and get in line to pay for it. You're really trying to hurry because you're already late for an appointment. Then, a woman with a shopping cart full of groceries cuts in line in front of you.
> *Woman:* "You don't mind if I cut in here do you? I'm in a hurry."
>
> Scene 2
>
> *Narrator:* You go to a ballgame with a reserved-seat ticket. When you arrive you find that a woman has put her coat in the seat for

which you have a reserved seat ticket. You ask her to remove her coat, and she tells you that she is saving the seat for a friend.
Woman: "I'm sorry, this seat is saved."

Scene 3

Narrator: You're in a restaurant with some friends. You order a steak, very rare. The waitress comes over to your table and serves you a steak which is so well done it looks burned. You really like your steak rare.
Waitress: "I hope you enjoy your dinner, sir." (Eisler, Miller, & Hersen, 1973, p. 297)

Client responses to such scenes are recorded on audiotape or videotape. They are scored along dimensions such as: (1) latency of response; (2) duration of response; (3) loudness of voice; (4) affect or liveliness of voice; (5) refusal to comply with unreasonable request; and (6) overall assertiveness.

Role playing is widely used to measure social skills such as assertiveness (Rathus, 1972); ability to express anger (Wagner, 1968); and competence in personal problem-solving (Arnkoff & Stewart, 1975).

Simulations

The laboratory environment can be structured to closely approximate conditions in the natural environment. The degree of realism achieved depends to a great degree on the ingenuity of the clinician.

In one case report (Lubetkin, 1975), a planetarium served as a laboratory for observation of a female patient who complained of an overwhelming fear of thunder and lightning. Assessment and treatment procedures were conducted in the planetarium, while a 3-minute showing of a thunderstorm was played. The simulated storm was complete with realistic visual and auditory cues. The patient watched the "storm" many times and learned to deal with her anxiety in the planetarium. Following this, she was able to deal with and overcome anxiety experienced during actual storms.

Realistic laboratory simulations are also used with flight phobics. Surwit (1975, personal communication) has employed mock-ups of passenger jet airliners in treating groups of patients who are afraid to fly. Bernstein and Beatty (1971) had a flight phobic sit in a Link aircraft trainer while motion, noise, and aircraft responses to a variety of flying conditions were simulated.

Realistic barrooms have been created in laboratories to study the behavior of chronic alcoholics (Nathan, Goldman, Lisman, &

Taylor, 1972; Schaefer, Sobell, & Mills, 1971). Realistic laboratory simulations are helpful aids in behavioral assessment. They are often rather difficult to engineer, however.

In sum, the laboratory is a convenient setting for assessment. It is also a setting where behavior can be precisely measured and recorded. A greater degree of environmental control is possible in the lab than exists in self-observation or naturalistic observation. The clinician can elicit behaviors that rarely occur in the natural environment because the client avoids situations calling for these behaviors. For example, the unassertive patient avoids many social situations. In the laboratory, he can be confronted with the very interpersonal situations he avoids and can be observed as he attempts to deal with them.

Issues in the Use of Observations

It is widely assumed that one of the major characteristics that differentiates behavior therapy from other types of therapy is an emphasis on the systematic collection of objective data through direct observation. The results of a review (Keefe & Sirota, 1977) of assessment methods employed with adult outpatients, however, raise serious questions about this assumption. Studies published in leading behavior therapy journals were rated by independent judges. The majority of studies relied on subjective assessment methods, such as self-observations or questionnaires. Fewer studies used more objective measures, such as direct observation in the naturalistic setting or laboratory setting. The obvious question is: "Why are more objective methods of assessment seldom used with adult outpatients?" The answer becomes apparent if one considers some of the practical and methodological problems in the observation of adult outpatients.

Practical Issues

The costs of observations are often prohibitive. In naturalistic observations, the major cost is in time. In order to observe a client in his natural setting, the therapist or observer often has to travel a considerable distance. Simply finding the client's house on a home visit can prove to be no easy task. Those who have conducted home visits can attest to the frustrations of locating a client's residence in a maze of housing project streets or in a rural area where there are few, if any, street signs.

In laboratory observations, there are the additional costs of space and equipment. This is particularly a problem for methods of laboratory observation which require expensive electromechanical programming and recording equipment—for example, automated interviews (Keefe & Webb, 1974); free operant systems (Nathan, Schneller, & Lindsley, 1964); and various electromechanical signaling systems (Katz, 1974; Thomas, Carter, & Gambrill, 1971). Advances in solid-state technology promise to reduce these costs in the future. At present, however, the cost of even relatively simple laboratory equipment, such as videotape recorders, is beyond the resources of most clinical facilities.

The relatively low cost of self-observation procedures accounts for their popularity with practicing clinicians. The cost for materials for self-observation, such as notebooks and counters, is minimal. The costs in terms of time are borne by the patient himself.

Methodological Issues

There are a number of methodological problems involved in making observations (Lipinski & Nelson, 1974). One set of problems consists of procedural issues. What coding system is to be used? How much observation is enough to ensure representativeness of data? How is reliability to be calculated? In actual practice, these procedural questions usually are answered before observation is begun. The complexity of coding systems varies depending on the nature of the presenting problem and availability of trained observers. The number of observations depends mostly on practical factors. The clinician chooses one or another method of calculating reliability, based on his purposes, the nature and scope of the presenting problem, and the time span covered (Johnson & Bolstad, 1973).

A second set of methodological problems stems from the reactive effects of the observation process. When a client attends to his own behavior, the observation process itself produces changes in behavior (Kazdin, 1974; Kopel & Arkowitz, 1974). Changes in behavior often occur in the desired therapeutic direction. For example, Ernst (1973) found a dramatic reduction in lip biting when a client was simply instructed to keep track of the number of times he bit his lip each day. Reductions in frequency of smoking (McFall, 1970; McFall & Hammen, 1972) and obsessive cancer thoughts (Frederiksen, 1975) have also occurred during self-recording. Reactive effects also occur when behavior is measured by independent observers.

A third set of problems is due to observer bias. An observer's report is biased when it is affected by factors other than the mere occurrence of particular target behaviors. Factors inherent in the observer, such as expectation of results or knowledge that reliability checks will be made (Reid, 1970) influence the data collected (Rosenthal, 1963, 1966).

The effects of reactivity and observer bias can be reduced if several steps are taken by the clinician (Johnson & Bolstad, 1973). First, clients should be provided with a thorough explanation of the reasons for observation. This may lessen defensiveness and anxiety about the procedures. Second, observations can be conducted over long time periods so that clients habituate to the novelty of the process. Third, whenever possible, less obtrusive recording techniques, such as automatic recording devices, should be employed.

In summary, a variety of observation strategies are available for the assessment of adult outpatients. Although direct observations in the naturalistic or laboratory setting may be the most scientifically sound strategy, these methods are often impractical. Behavioral clinicians tend to rely heavily on self-observation procedures.

Psychophysiological Analysis

Patients often describe their behavior using such terms as: "I felt as though I had butterflies in my stomach"; "My heart was racing"; or "I felt tense." Psychophysiological analysis provides the clinician with a method of directly measuring some physiological correlates of such verbal reports. This type of assessment is important to the behavior therapist because physiological responses often play a central role in the development and maintenance of maladaptive behavior patterns.

Measurement of Physiological Responses

Basic to the process of psychophysiological analysis is the measurement of physiological responses. Physiological measurement systems are composed of three basic components (Brown, 1972). The first component of the system, the *transducer,* is attached directly to the subject and serves an input function—it picks up a signal. Transducers vary in form depending on the response characteristics being monitored. For example, thermistors measure changes in skin temperature while electrodes measure changes in bioelectric activity either directly (as in electromyography) or indirectly (as in the measurement of skin resistance). The second component, the *signal-processing unit,* amplifies, filters, and modifies the signal. It functions as a throughput. The third component of the system, the *output display,* transforms electrical signals to a form that can be conveniently observed, such as the deflection of a pen on a polygraph chart, or numbers on a digital readout.

Physiological response measures are typically taken in a laboratory consisting of two adjacent rooms. In one room, the subject sits in a comfortable chair with transducers attached to him. This room is constructed in such a way that precise control of environmental variables—for example, temperature, humidity, illumination, and noise level—is possible. A second room contains a polygraph and programming equipment. Although the polygraph is an imposing piece of equipment, it houses devices which perform two rather straightforward functions: throughput and output. The polygraph is flexible and can perform throughput and output functions simultaneously for two or more response systems: for example, heart rate, respiration, and muscle activity can be monitored concurrently.

Uses of Psychophysiological Analysis

Psychophysiological analysis is used by behavioral clinicians in several ways. One use involves taking physiological measures before and after patients engage in problematic behaviors. Pulse rates of obsessive-compulsive patients, for example, are taken before and after touching "contaminated" objects or engaging in ritualistic hand washing (Hodgson & Rachman, 1972). A second use is in measuring the physiological accompaniments of certain cognitive images. The effects of phobic imagery (Lang, Melamed, & Hart, 1970) and obsessive imagery (Rabavilas & Boulougouris, 1974) can be assessed using this procedure. A third use of psychophysiological analysis involves measuring the effects of brief presentations of certain stimuli on physiological responding. Rates of habituation to these stimuli may then be studied. Patients having a higher background level of physiological arousal (for example, anxiety neurotics) often show slower rates of habituation (Lang, 1971). Watson, Gaind, & Marks (1972) measured the heart rate of a patient with a profound fear of cats over repeated presentation of phobic stimuli. Early in the session, looking at either a black or ginger cat from 6 feet produced marked increases in heart rate in the patient, which habituated with prolonged exposure. A similar increase in heart rate occurred when the patient first touched one of the cats, but this too habituated. By the end of the session, only slight changes in heart rate occurred, even when a cat happened to touch the patient's leg. Reductions in physiological responsivity were associated with cognitive and behavioral improvement.

A fourth use of psychophysiological analysis is to monitor the physiological effects of ongoing therapeutic interventions, such as aversion therapy (Gaupp, Stern, & Ratliff, 1971) or systematic or in vivo desensitization (Agras, 1967; Baker, Cohen, & Saunders, 1973; Borkovec, 1974; Wolpe & Flood, 1970).

The possibility that physiological self-control can be learned has generated a great deal of interest both in laymen (Stoler, 1974) and in scientists (Schwartz, 1973). Biofeedback is one of the ways in which physiological self-control can be learned. In biofeedback sessions, visual or audio displays transmit accurate information to the subject regarding changes in a physiological response. The basic tenet underlying this procedure is that subjects can use the information provided by feedback displays to voluntarily control their physiological responses.

Biofeedback techniques initally were used only in laboratory settings. More recently, clinicians have begun to employ portable, self-contained biofeedback measurement systems in office or home settings. Biofeedback units can be assembled by the practitioner (Yonovitz & Kumar, 1972; Leaf & Gaarder, 1971) or purchased from commercial suppliers (Rugh & Schwitzgebel, 1975a, 1975b). Successful clinical applications of biofeedback training have been reported in the treatment of a variety of disorders, including hypertension, epilepsy, cardiac arrhythmias, tension headache, migraine headache, Raynaud's disease, asthma, and various neuromuscular disorders (Blanchard & Young, 1974). At present, however, the evidence for the effectiveness of biofeedback is tentative, and considerable research needs to be conducted before this method becomes a viable clinical tool. Some of the issues which need to be addressed in this research are: the importance of patient motivation; relevant patient characteristics; the necessity for training generalizations; and cost-effectiveness relative to other treatment procedures (Schwartz, 1973; Shapiro & Surwit, 1976).

Issues in the Use of Psychophysiological Analysis

The interpretation of data gathered through psychophysiological analysis is not always a simple process (Lang, 1971). The meaning of an increase in heart rate or a drop in blood pressure may not be entirely clear. An increase in heart rate can not simply be interpreted to mean that there has been an increase in anxiety. Lang (1971) has referred to this tendency as the *indicant fallacy:*

> Nearly all physiological responses can be generated by a great variety of internal and external stimuli, and it seems unlikely that any physiological event could be used in an exact substitutive way as an index of psychological state. Thus, by observing the physiology of an organism, we are not able to go backwards and reconstruct the stimulus input or the psychological state that contributes to its generation. (p. 89)

Psychophysiologists argue that the meaning of a particular response is most clearly interpreted when several responses, such as heart rate, respiration, and skin conductance, are measured at once. Patterns of change in single and multiple responses may then be examined and compared. The clinician should be cautious in interpreting data gathered using a single physiological response, such as are provided by many biofeedback units.

Psychophysiological analysis also involves complex technical and methodological considerations (Lacey, 1959). The clinician needs to be able to rule out artifacts in order to obtain valid data. As Lang (1971) has pointed out, this may require background and training in diverse fields such as psychology, general physiology, computer science, and electronics. Few clinicians have such a background.

Behavior therapists only recently have begun to use psychophysiological methods to assess patients in clinical settings. The development of practical, inexpensive, self-contained recording systems has been a major impetus in this process. The main advantage of psychophysiological analysis is that it provides a direct measure of physiological responding. The major disadvantages of psychophysiological analysis are: (1) it is difficult to interpret the meaning of changes in physiological responses; and (2) the technical problems inherent in psychophysiological analysis may be beyond the grasp of most clinicians. Psychophysiological analysis is a behavioral assessment method that is currently used by few practicing clinicians, but it is likely to become more widely used in the near future.

In Chapter 4, we considered methods commonly used in the assessment of adult outpatients: (1) the interview; (2) questionnaires; (3) observations; and (4) psychophysiological analysis. We found that each method had its inherent advantages and disadvantages. In practice, the behavioral clinician seldom relies on only one method. Typically, a combination of methods is employed. Chapters 5 and 6 illustrate this process.

CHAPTER 5

Behavioral Assessment of Marital Discord

In Chapter 4, we reviewed methods of behavioral assessment for the individual adult outpatient. In Chapter 5, we focus on the assessment of relationship problems between adults in a marriage. Although marital discord or distress is the topic of concern, the approach and procedures outlined are applicable to adult relationship difficulties in general.

The application of the behavioral approach to the treatment of marital discord is a recent development which is growing rapidly. Much of the behavioral work on marital discord has been based upon a model which emphasizes the couple's faulty behavior-change strategies, including difficulties in communication and problem-solving (Jacobson & Martin, 1976; Stuart, 1969; Weiss, Hops, & Patterson, 1973). We will describe all five stages of behavioral assessment of marital discord, with particular focus on the unique aspects of relationship problems.

Problem Identification

The Initial Interview

A major problem facing any marital therapist is determining the commitment of the couple to each other and to therapy. It is likely that each spouse will enter the intake session with a desire to tell his or her part of the story. The agenda often is hidden. One agenda is the "day in court." For example, the husband may be out to *prove* that he was right and his wife was wrong. The therapist is seen as a judge who pronounces a verdict, "yea" or "nay." A second agenda is the "doomsday" one. Husband, wife, or both may be motivated to show that *nothing* the therapist suggests will work. Therapy may thereby be used to rationalize or legitimize an already entrenched decision to divorce or separate.

The issue of commitment to treatment is raised early in the interview. If there appears to be a lack of commitment, the therapy contract is renegotiated. If commitment appears adequate (i.e., the couple reports they are committed, after discussion and observation support this), assessment can continue.

Initially permitting the couple a brief period of unstructured interaction provides the interviewer with a sample of their interactive style. The practitioner who moves too quickly into a structured behavioral interview may miss this important sample of behavior. In fact, a major part of the intake interview is often spent listening to the couple's complaints about each other, which they will likely spontaneously provide. Their descriptions may be inexact, exaggerated, and quite emotional. The interviewer takes notes so that he can attempt to operationalize these vague complaints at a later point.

Peterson (1977, pp. 305–316) has suggested the following outline to structure the clinical marital interview:

1. How the relationship began
2. Important changes during the course of the relationship
3. Perceived affection and authority
4. The major problems in the marriage
5. Recurrent sequences of problematic interaction
    a. Detailed descriptions of the occasions
    b. The setting and context
    c. What was done and said
    d. Feelings about each other
    e. How they influenced each other
    f. Outcome
    g. Whether each spouse felt any injustice was done by omission or commission (contract violation)
6. Enjoyable experiences with detailed examples and analysis as in #5.

The progression in the interview is from the more general and global to more specific detailing. The following transcript illustrates the interviewer's attempts to derive behavioral referents for vague complaints:

THERAPIST: You mentioned that you felt your husband doesn't love you any more. Can you describe why you feel that way?

WIFE: He just doesn't show he cares about me. It's mainly his attitude.

THERAPIST: Can you give me some examples during this past week or two?

WIFE: Definitely! He goes out bowling and plays cards with the guys, but doesn't have time for me.

THERAPIST: What kinds of activities would you like to do together?

WIFE: I would be satisfied if he just talked to me in the morning instead of reading the paper.

THERAPIST: So what you are saying is that you want him to pay more attention to you and interact more on a daily basis.

WIFE: Yes. But, I'd also like him to take me out once in a while.

THERAPIST: How often do you go out together now?

WIFE: Maybe twice a month, but it's usually with the kids to a restaurant.

THERAPIST: What types of places would you like to go to together?

The interviewer should point out the importance of the clarification he is seeking, thus helping the couple to shift their descriptions to specific, more precise descriptions. At the conclusion of the intake session, the interviewer typically asks the couple to complete questionnaires at home and return them by mail before the next scheduled session.

Questionnaires

Early assessment activities set the stage for subsequent intervention by establishing norms, expectations, and client-therapist relationships. Behaviorally oriented questionnaires represent an active step in behavioral marital therapy beyond data collection. They can teach the couple new ways of describing or thinking about their relationship. Couples often remark that the questionnaires help point out specific positive aspects of the relationship they had not been aware of, e.g., child-rearing practices.

One of the potentially most useful questionnaires is the Marital Pre-Counseling Inventory (Stuart & Stuart, 1972). This comprehensive questionnaire includes sections on: (1) family composition; (2) spouses' activities for a typical week; (3) spouses' pleasing behaviors toward each other and positive behaviors to be accelerated; (4) individual resources, goals, interests, and mutual interests; (5) perceptions of current responsibility for decisions in major areas and indications of ideal division of responsibilities; (6) satisfaction with the way major areas are handled in the marriage; (7) spouse communication; (8) sexual satisfactions; (9) child-rearing; (10) commitment and optimism; and (11) other positive changes desired. The Marital Pre-Counseling Inventory is 13 pages long, requiring considerable time and effort to complete; however, we consider it worth the effort as the information gathered is quite valuable. The Marital Pre-Counseling Inventory was developed within a behavioral clinical approach to marital problems, so the information collected fits well into a behavioral, S-O-R-K-C conceptualization. The format aids in identifying specific behavioral targets for further measurement and intervention. Furthermore, the inventory is a useful learning tool to teach the couple the language and concepts used in a behavioral approach. Finally, the inventory examines the context of the relationship in a holistic manner in terms of perceived roles, expectations, and spouses' daily routines.

A second paper-and-pencil measure of marital satisfaction is the Locke-Wallace (1959) Marital Adjustment Scale (MAS). This test provides a good global self-report evaluation of marital satisfaction. It is composed of 15 multiple-choice questions dealing with various aspects of the marital relationship. A copy of the test, along with the score for each response, appears as Table 5.1. A total score is derived by simply adding the scores obtained from each of the 15 items. Low scores indicate more distressed relationships. A cutoff of 100 has been viewed as a good discriminator between stable and unstable marriages. Adequate reliability and validity have been reported for the MAS (Locke & Wallace, 1959). The inventory requires less than 5 minutes for administration and scoring. Since the spouses complete the form independently, large differences between or basic similarities in the total scores for the partners provide additional assessment information.

The MAS has the advantage of providing an "economical" quantified measure of overall marital satisfaction. It can be given before and after treatment to easily assess changes in satisfaction levels. We use both the Marital Pre-Counseling Inventory and the MAS to supplement each other so that we obtain specific information consistent with the behavioral approach as well as an overall index of the relationship.

By receiving the questionnaires before the next scheduled session, the clinician can review the responses and select areas for

Table 5.1
The Marital Adjustment Scale[1]

1. Check the dot on the scale below which best describes the degree of happiness, everything considered, of your present marriage. The middle point, "happy," represents the degree of happiness which most people get from marriage, and the scale gradually ranges on one side to those few who are very unhappy in marriage and on the other to those few who experience extreme joy or felicity in marriage.

0	2	7	15	20	25	35
.	.	.	.	.	.	.
Very Unhappy			Happy			Perfectly Happy

State the approximate extent of agreement or disagreement between you and your mate on the following items. Please check each item.

	Always Agree	Almost Always Agree	Occasionally Disagree	Frequently Disagree	Almost Always Disagree	Always Disagree
2. Handling family finances	5	4	3	2	1	0
3. Matters of recreation	5	4	3	2	1	0
4. Demostrations of affection	8	6	4	2	1	0
5. Friends	5	4	3	2	1	0
6. Sex relations	15	12	9	4	1	0
7. Conventionality (right, good, or proper conduct)	5	4	3	2	1	0
8. Philosophy of life	5	4	3	2	1	0
9. Ways of dealing with in-laws	5	4	3	2	1	0

10. When disagreements arise, they usually result in:
    Husband giving in=0   Wife giving in=2
    Agreement by mutual give-and-take=10
11. Do you and your mate engage in outside interests together?
    All of them=10   Some of them=8   Very few of them=3
    None of them=0
12. In leisure time, do you generally prefer: to be "on the go," or to stay at home? Does your mate generally prefer: to be "on the go," or to stay at home?
    Both stay=10   Both go=3   Disagree=2
13. Do you ever wish you had not married?
    Frequently=0   Occasionally=3   Rarely=8   Never=15
14. If you had your life to live over, do you think you would:
    Marry the same person=15   Marry a different person=0
    Not marry at all=1
15. Do you confide in your mate?
    Almost never=0   Rarely=2   In most things=10
    In everything=10

Reprinted by permission of Karl M. Wallace and H. J. Locke.

[1] The numbers appearing on this scale beside each response to an item are for scoring purposes only. They should be deleted from the scale prior to administration.

further investigation in subsequent interviews. Although the couple may easily overlook the positive aspects of the relationship, the clinician should take care to specifically assess the strengths and positive experiences of the marriage as well as the problem areas. The information gathered from the questionnaires, the global and behavioral interviews, and actual samples of the couple's in-office interactions provide a comprehensive initial description of marital dysfunction.

Measurement and Functional Analysis

There are a number of behavioral models of marital discord which are useful in guiding the therapist to focus on various classes of behavior for measurement and functional analysis (e.g., Stuart, 1975).

A Behavioral Model of Marital Discord

One useful classification system has been developed by Weiss, Patterson, and their colleagues at the University of Oregon (see, e.g., Birchler, 1972; Patterson, Weiss, & Hops, 1975; Vincent, Weiss, & Birchler, 1975; Weiss, Hops, & Patterson, 1973; Weiss & Margolin, 1975). This model focuses on three classes of behaviors: (1) affectionate exchanges; (2) problem-solving behaviors; and (3) behavior change attempts. The Weiss-Patterson conceptualization of marital discord maintains that marital problems primarily stem from faulty behavior change operations. In specifying the three major classes of behavior to focus upon, the model implies that the couple's style or process of dealing with conflict, differences, or problems is the primary area of concern. The content of marital issues is not ignored, but is rather seen as something through which basic problems in the three classes of behavior occur.

Affectionate Exchanges .

This category includes behaviors ranging from intimate physical or sexual exchanges to gifts for the spouse or exchanging positive verbal statements of praise, acknowledgment, or compliments. In layman's terms it consists of "being nice to each other." This class of behaviors covers daily exchanges such as a hello–good-bye kiss, as well as special exchanges such as a weekend away without the children.

There is evidence that distressed couples tend to exchange affectionate behaviors at relatively low rates when compared to nondistressed couples (Vincent et al., 1975). These low rates of positive exchange (and relatively high rates of negative exchange) suggest a relationship deprived of positive experiences. The low rates of positive exchange may be due to a strong negative "motivational" state between the spouses or to interpersonal skill deficits of one or both partners: one or both may not want to exchange affectionate behaviors or may not know how to do so competently.

Problem-Solving Behaviors

Distressed marital couples are notorious for their ability to start discussing a specific problem and to sidetrack to tangential problems, historical issues of conflict, or total irrelevancies. For example, a discussion of respective roles in the marriage can easily deteriorate to blaming in-laws for their "poor job in raising their child." Such discussions can obviously become vicious and destructive while moving further and further from any resolution of the original issue. Deficits in communication skills, such as relying on "mind-reading" for making assumptions about the other partner's expectations or motives, may further contribute to poor problem-solving. In general, difficulties arise in this class of behavior because the couple is ineffective in one or more of the following components of problem-solving: (1) specifying or pinpointing the components of the problem; (2) clarifying what specific changes they desire; (3) brainstorming or generating potential solutions; and (4) arriving at a mutually agreeable solution.

Behavior Change Attempts

Distressed couples often fail to adequately apply principles of positive reinforcement and shaping. More often they use elements of aversive control or negative reinforcement. For example, a spouse nags or complains until the other finally "gives in" and takes out the garbage. The aversive strategy of nagging, complaining, or plain "bitching" is reinforced by the occurrence of the desired change in the partner's behavior. If the partner resists change until extremely high levels of aversive behavior are emitted by the spouse, the "complainer" is shaped into using highly obnoxious behaviors. There is ample evidence to suggest that negative control begets negative reciprocation (Patterson & Hops, 1972). Couples often employ *reciprocal coercion* as their primary strategy for behavior change. This leads to "resentment, frustration, hostility and often outright aggression" (Vincent et al., 1975).

The Weiss-Patterson model of marital discord integrates concepts from social psychology (reciprocity), systems theory, communication theory, decision theory, and social learning theory. In so doing, it permits an empirically oriented description of traditional concepts of marriage. For example, *trust* is viewed as an expectation of receiving something positive when one gives something positive—reciprocal positive exchange (see, e.g., Thibaut & Kelley, 1959). The model does not ignore the couple's traditional concepts of marriage. Rather, it attempts to relate these concepts, love, trust, roles, etc., to specific behaviors, cognitions, feeling states, and environmental factors. The three major categories of behavior that are emphasized guide the clinician through a S-O-R-K-C analysis particularly relevant for the problem of marital distress.

The Weiss-Patterson model is a very useful framework for behavioral assessment and therapy of marital discord. It was selected for summary presentation in this chapter because it appears to be comprehensive, quite useful, and more clearly linked to research findings than competing models (cf. Jacobson, 1977; Jacobson & Martin, 1976). As a cautionary note, however, the reader is reminded that the field of marital behavior therapy is in its nascent stage of development. Thus, the therapist should not bind himself to any one system at this time, since the field is under constant development.

Methods of Measurement and Functional Analysis

Once marital problems of style and content are identified, the focus shifts to measurement. Behavioral measurement procedures that are available include self-recordings, laboratory observations, and naturalistic observations in the couple's home. In large research projects (e.g., Peterson, 1976; Weiss et al., 1973), distressed couples have received a battery of behavioral assessment procedures which have included all three types of measures. Practical constraints, however, limit the options typically available to the practicing clinician and some choices are thus necessary.

Self-Recordings

Self-recording procedures are helpful in gathering detailed information on the rates and types of positive and negative exchange, as well as measurement of specific target behaviors. Behaviors recorded may include, for example, touching, complaining, paying compliments, kissing, or demanding statements. The spouses may be asked to monitor specific behaviors or more general classes of behaviors. A more demanding assessment task requires self-recording of all perceived positive and negative exchanges. The burden of this task is eased by limiting these recordings to specific time periods and/or providing the couple with wrist counters. Jacobson (1977) reports the use of self-recording to assess daily time spent talking between partners, time spent by husband participating in housework, and time spent by husband interacting with children.

Laboratory Observations

Laboratory analogue or simulation tests provide a second method of more detailed measurement. One such analogue reported in the literature is to ask the couple to discuss a specific issue of conflict for 10 minutes while attempting to reach a resolution within that time (see, e.g., Weiss et al., 1973). The therapist helps the couple to choose the topic to discuss beforehand to assess problem-solving skills over the range of minor and major problems. The therapist can ask the couple to select two problem areas for two separate conflict-resolution attempts. One problem should be viewed by both partners as major (e.g., handling of finances) and the other as relatively minor (e.g., weekend recreation). These interactions should occur without the presence of the therapist and should be audiotaped or videotaped for subsequent analysis. In these analogue situations, couples commonly get into heated arguments typical of those occurring at home.

Analyses of interactions occurring during the conflict-resolution task can be made by examining the following classes of behaviors:

1. **Behavior description**

    a. Statements pinpointing, objectifying, or operationalizing the problem (e.g., "You leave your clothes on the dresser" in contrast to a vague statement such as "You are a messy person").
    b. Sidetracking—bringing up irrelevant points or issues, especially history of the problem or other conflicts (e.g., "Since we were married you've never hung up your clothes," or "Talking about the closet, when are you going to put up those shelves?").
    c. Tracking—returning to relevant issues.
    d. "Mindreading"—statements assuming the motivation, intent, or attitude of the other (e.g., "I know you are saying you'll put

your clothes away, but I know what you are really thinking," or "You're just going along because you want to appease me, but you don't really care how the room looks").

2. Problem analysis and resolution

a. Statements of what specific changes would solve the problem, with qualifying details (e.g., "I would be satisfied if you would put your socks and underwear in the hamper and hang up your pants so I wouldn't have to iron them so often. I don't mind if you leave your shoes by the bed and put the permanent press shirts on the back of the chair, as long as they're put away next morning").
b. Suggesting trade-offs, compromises, or contracts.
c. Generation of alternative potential solutions.
d. Reaching a mutually agreeable resolution which is realistic.

3. General communication skills of support or understanding

a. Paraphrasing—reflection.
b. Verbal-nonverbal incongruities.
c. Eye contact.
d. Acknowledgment or agreement when appropriate.
e. Balance of talk time between partners.

4. Exchanges of positives and negatives

a. Use of reinforcement—verbal and nonverbal.
b. Use of punishment—verbal and nonverbal.
c. Use of extinction—ignoring.

There are several procedures available for analyzing or scoring the interactions of the conflict-resolution task. One approach is to employ a highly formalized, complex coding system to analyze in detail the videotape recordings of every response by each spouse. This procedure has been employed by several research investigators (e.g., Birchler, 1972; Vincent, 1972; Vincent et al., 1975; Jacobson, 1977). The major problem with this approach is that it requires highly trained observers to rate the tapes and costly videotape equipment. To most clinicians, these requirements are well beyond the resources available. A second approach is to review audiotapes of the couple's interactions to describe the general "flavor" of the interaction across the major classes of behavior which are of interest. While such descriptions may provide some clues for the behavioral clinician and are practical, they are not very precise. A third alternative is available which is more precise and also is practical for the clinician: to take frequency counts of specific responses or classes of behaviors. A sample data sheet for this purpose appears as Table 5.2, listing responses that are useful to measure. The clinician may take frequency counts of only one or two of the responses, or he may wish to measure a number of responses across behavioral categories. In the latter case, the tape can be reviewed many times, while the rater counts the frequency of only a few target responses each time.

Naturalistic Observations

Naturalistic observations are made in the settings in which the couple actually is having their problems. Such measurements can easily be obtained from the couple's home environment using audiotapes of select "critical" time periods when problems are most likely to occur (e.g., dinner discussions). The couple can also be instructed to turn on the tape recorder when a conflict arises; this procedure also can be used when the interaction is very positive, so that the therapist can sample those events as well. The audiotape recordings are typically made on a number of days rather than being taped on only one occasion, thus providing time for the couple to overcome the common initial reactive effects of tape recording in the home. As in the case of laboratory tape recordings, the tapes from the home environment can be analyzed via the three methods already discussed.

It is practical for the clinician to collect measurements across self-recordings of target behaviors in the naturalistic environment, tapes of laboratory simulations, and tapes of interactions in the home. Data from these sources integrated with the interviews, as well as paper-and-pencil instruments, provide a comprehensive, multiple behavioral assessment for case formulation and treatment plans.

Matching Treatment to Clients

The first two stages of behavioral assessment permit both the couple and the therapist to judge whether or not the behavioral approach in general is appropriate for them. The active participation and responsibility required of the couple in the assessment procedures can be viewed as a "test" of their acceptance of a treatment approach which will have similar requirements. If one

Table 5.2
Data Sheet for Observation of Problem-Solving Task

Couple _____
Date _____ Coder _____

Major _____
Problem selected _____ Minor _____

Instructions: For each occasion that a behavior occurs, place a check (✔)
under the column of the spouse who emitted the response. Listen to or
observe the tape once for each of the five major behavior categories.
Comments may be added to describe style for each category.

	Husband	Wife

I. *Problem identification*
   1. Pinpoints
   2. Operationalizes
   3. Sidetracks
   4. Tracking
   5. "Mindreading"

Description of style:

II. *Problem analysis and resolution*
   6. Specification of desired outcome
   7. Qualifying conditions
   8. Alternative solutions
   9. Trade-offs or compromises
   10. Mutually agreeable solution

Description of style:

III. *Communication skills*
   11. Paraphrasing—Reflection
   12. Verbal-nonverbal inconsistencies
   13. Eye contact
   14. Acknowledgment/agreement
   15. Percent talk time

Description of style:

Table 5.2 (cont.)

IV. Exchanges of positives and negatives
  16. Positives (reinforcements)
     a. Verbal
     b. Nonverbal
  17. Negatives (punishments)
     a. Verbal
     b. Nonverbal
  18. Ignoring (extinction)

Description of style:

Ask each spouse the following questions immediately after the task:

1. How representative of the usual situation was that conflict resolution?

Husband:	1	4	7
	Not at all	Somewhat	Extremely
Wife:	1	4	7
	Not at all	Somewhat	Extremely

2. Is it usually more positive or more negative?

Husband:	1	4	7
	Much more positive	About the same	Much more negative
Wife:	1	4	7
	Much more positive	About the same	Much more negative

or both partner's expectations are too discrepant from the behavioral approach, or if their resentment toward each other is too great, they would likely have withdrawn from a therapy program with such requirements by the third stage.

Obviously, the therapist does not wait until the measurement phase has been completed before sharing his basic approach to marital discord with the couple. In order to justify and explain the procedures of assessment used, the therapist already has prepared the couple by explaining the basis for the behavioral approach in understandable terms and concepts. At the completion of the measurement and functional analysis stage, the therapist holds a sharing conference with the couple to summarize the information he has collected on their relationship in a behavioral case formulation. Problematic content areas are listed as well as problems in the partners' style of handling their differences. Targets for intervention are delineated as part of a discussion of mutual goals. The therapist and couple must decide which problem area will be the first focus of treatment.

There are distinct advantages to the therapist's suggesting that the couple begin with a relatively minor conflict area. This strategy should increase the likelihood of initial success and give the couple practice in applying the behavioral approach before tackling more difficult problem areas. Emphasis is placed upon their learning the behavioral principles and applying them to attain new patterns of behavior which will resolve their current problems. Long-term goals also are typically discussed in terms of learning how to prevent future problems and maintain the improvement in the relationship.

The multiple sources of data already collected provide sufficient information on deficits, excesses, and inappropriate response patterns to guide the clinician in his choice of interventions.

A variety of intervention procedures or "modules" are used in the behavioral treatment of marital discord. Weiss and his co-workers (1973) review four comprehensive intervention training modules which are based upon their model of marital discord: (1) discrimination and reinforcement training for "pleases" and "displeases"; (2) problem-identification training; (3) communication training; and (4) contract and negotiation training. These four training programs cover the categories of affectionate exchanges, problem-solving behaviors, and behavior change attempts. The assessment information permits the clinician to identify which of the training programs is relevant, as well as which components of each program are needed. (The reader who is unfamiliar with these intervention procedures is referred to the references given in this chapter for detailed discussions.) For example, a given couple may more than adequately recognize and exchange positive and affectionate behaviors. Their ability to identify problems and communicate in general may be proficient. However, the assessment information may clearly point out deficits in problem-solving abilities; thus, a focus upon problem-solving skills, contracting, and negotiating would be indicated. Since particular problem or conflict areas or issues have also been identified in assessment, the intervention training program(s) for this couple can be designed to work through these particular problem areas in a selected order.

Assessment of Ongoing Therapy

When working with distressed couples, the assessment procedures of earlier stages often are used as vehicles for current treatment and assessment. For example, self-recording of positive exchanges at home has been reported as an integral component of a program in discrimination and reinforcement training (Weiss et al., 1973). A couple is instructed to monitor the number of positive exchanges ("pleases") given and received by each other on a daily basis for a period of time such as a week. One specific intervention procedure here is to instruct one or both spouses to select one of these days as a "love day," to double or triple the rate of positives given to the partner. The partner is not told which day has been selected. The self-recording data is reviewed to assess whether the procedure was implemented and whether it was perceived by the partner. The effects of increasing positive exchanges are discussed with the couple, with emphasis on the discrimination and reinforcing aspects and the effect on their feelings and attitudes toward each other. Graphs of the daily rates of positive exchange are useful tools to assess and demonstrate the effect of the "love day" intervention procedure. In general, self-recording by the spouses is a useful strategy to assess specific changes in behavioral targets as intervention programs continue. For clinical purposes, the possible reactive effects of such monitoring may aid in initial behavior change which is then maintained by positive reinforcement and general positive changes in the couple's style of interacting.

Naturalistic observation and in-office simulation procedures via audiotape recordings may be used both to train and to assess communication, problem-solving, or behavior change intervention training programs. For example, the couple is asked to tape a home discussion of a selected problem area. The tape is then reviewed in the office session by the therapist, who points out problematic and constructive interactions. The therapist provides feedback, and models more appropriate statements. The couple is then instructed to discuss the same problem area, incorporating the suggestions reviewed. The audiotape of this second attempt to arrive at a resolution can then be further reviewed and analyzed by the couple with therapist input. The tapes permit assessment followed by the couple's corrective practicing. Assigning subsequent taped home sessions of conflict-resolution in different problem areas provides direct assessment of skill training and also permits therapist feedback and further opportunities for in-office corrective practice.

The information collected by the assessment procedures allows the clinician to evaluate the effectiveness of the intervention training programs on an ongoing basis. If the couple is not applying the learning principles adequately, the therapist can decide to provide more structured training and/or shift the focus to more basic behavioral skills. An important feature of the procedure is that the therapist is using a monitoring/data collection system which directly influences the therapy process. An additional clinical advantage accrues from the assessment of ongoing therapy. As positive changes occur in the relationship, the assessment procedures highlight the improvements. Often couples will make strong positive inferences or conclusions about the changes that have occurred because the assessment procedures have focused their attention on these positive changes. This self-perception effect for the couple may play an important role in building upon initial small positive behavior changes.

Evaluation of Therapy

The behavioral treatment of marital discord continues until the couple is satisfied that they have changed their behavior in the desired direction and are able to maintain that change. The goal of therapy is for the couple to demonstrate they can actively use behavioral principles in their natural environment on a regular basis before therapy is terminated.

A complete evaluation usually follows a period of several weeks or months without therapy sessions or home assignments. The couple then returns for a follow-up session. The purpose is to discuss: (1) events that have transpired in the interim; (2) the couple's satisfaction with treatment and changes; (3) dealing with "partial relapses to old patterns"; and (4) unresolved conflict areas. Detailed assessment of these issues provides some of the best indicators of long-term marital improvement. This assessment also identifies targets requiring further intervention. When therapist and couple determine that long-term changes in behavior have been achieved, therapy is terminated. At this time, certain measures (such as questionnaires or a laboratory observation) are readministered. The results are summarized and shared with the couple to demonstrate changes occurring over the course of treatment.

CHAPTER 6

Behavioral Assessment of Sexual Dysfunction

The recent phenomenal growth in the practice of sex therapy in the United States leaves no doubt that a book on behavioral assessment would be incomplete if it did not cover this area. Treatment of sexual dysfunction clearly requires a thorough knowledge of human sexuality. Familiarity with the anatomy and physiology of sexual functioning as well as the influence of psychological and social factors is needed to understand the etiology, maintenance, and treatment of sexual problems. The focus of Chapter 6 is restricted to the assessment of psychological-social aspects of human sexual dysfunction. The interested reader is referred to excellent reviews of the biomedical aspects of sexual functioning by Masters and Johnson (1966) and Katchadourian and Lunde (1972).

A wide variation in the format of assessment and treatment for sexual dysfunction has been reported, including: (1) active physician involvement versus none; (2) mixed-sex co-therapist teams versus individual therapists; (3) a distant sex clinic versus local outpatient services; and (4) intensive programs (e.g., daily therapy sessions for two weeks) versus weekly sessions. The model of assessment outlined in Chapter 6 does not restrict itself to any of the above distinctions. That is best left to future empirical investigation and the resources available to the clinician. We will assume that the sexual dysfunction occurs within a relationship and that both partners will participate in the therapy program.

Problem Identification

"I'm impotent"; "I'm just not interested in sex"; "I'm frigid." Such statements are often heard as presenting complaints. The first task of assessment is to determine what the client means by

Table 6.1
Definitions of Commonly Applied Diagnostic Classifications for Sexual Dysfunctioning

Male Dysfunction

Erectile dysfunction
 Inability to obtain or maintain an erection for intromission and satisfactory intercourse
Ejaculatory incompetence
 Premature ejaculation
  Relatively quick ejaculation which precludes mutual sexual satisfaction
 Retarded ejaculation
  Delayed ejaculation which interferes with mutual sexual satisfaction
Low sexual drive
 Relatively minimal or no desire for intercourse or other heterosexual behavior

Female Dysfunction

Orgastic dysfunction
 Difficulty in achieving orgasm
Sexual unresponsiveness
 Difficulty in arousal
Dyspareunia
 Painful intercourse
Vaginismus
 Conditioned contraction of vaginal entrance muscles which prevents entrance of penis

---

these statements. For example, consider the client who complains of impotence: Has he ever had an erection? If he has, under what conditions? What does the wife mean who complains: "I'm not interested in sex." Is she ever aroused? Does this occur with a partner or through self-stimulation?

Early clinical and research investigations have led to a generally accepted classification system for sexual dysfunction. Table 6.1 presents the more commonly used categories and the generally accepted definitions.

Behaviorally oriented assessment offers a critically needed supplement to this classification system. In essence, the strategy of behavioral assessment permits the detailed analysis necessary to treat the individual case rather than the summary descriptor label. Presenting problems are tied to observable behaviors and environmental conditions. This task is not easy. The client may not be aware of certain behaviors, or may be uninterested in or embarrassed about talking about them. How are these difficulties in identifying presenting problems dealt with? Interviews and specially designed questionnaires are the primary assessment tools used for problem identification and clarification.

Interviews

The Initial Interview

Couples often enter the intake interview session with great apprehension. They may have never before discussed their highly emotional sexual difficulties with another person. They may be confused about what sex therapy actually is or is not. The couple may be embarrassed because they are not sure of the appropriate way of describing their problems.

To diminish these anxieties, we suggest structuring the initial interview with several introductory statements. Initially, the purpose and content of the first session is explained. The couple is told that the session is one of information-gathering for both the therapist and themselves. That is, the therapist will ask questions to identify problems they are having and he will also provide information about the sex therapy program. The clients are told that, based upon the information exchanged, there will be a mutual determination of whether this therapy approach "makes sense" for them. Before starting problem identification, the therapist further explains: "Most people find it somewhat difficult to describe their sexual problems and behaviors to someone who is basically a stranger to them." The couple is assured that they can use the terms they are most comfortable with in their descriptions. These preparatory statements will likely permit the therapist to proceed with problem identification within a less apprehensive context for the couple. Opening questions are general in nature, as the following dialogue suggests:

> **THERAPIST:** Why don't we start with one of you explaining what the problems are that led you to seek help?
>
> **HUSBAND:** We seem to be sexually incompatible.
>
> **THERAPIST:** What do you mean by that?
>
> **HUSBAND:** Well, usually when one of us is in the mood the other isn't and when we do have sex I usually come before my wife reaches her peak.
>
> **THERAPIST:** Mrs. ─────────, do you also see that as the major problem?

These initial questions allow each partner to describe the problem as he or she views it. Checking the perceptions of both partners will identify descrepancies which will require further exploration. If the couple basically agrees, the therapist has the benefit of a reliability check to increase confidence in the self-reports. This strategy of checking both partners' views is useful throughout assessment and therapy. In fact, discrepancies in information provided by the couple may indicate problems in perception or communication that are functionally related to the sexual problems.

As the interview progresses, the therapist asks more specific questions about the problems described, such as: "Does the erection problem occur as a loss of erection during foreplay or after entry? Or do you not achieve an erection at all? Are there times when you don't have a loss of erection? When did the problem first occur and were there major changes in your lives at about that time? What ways have you, or both of you, tried to deal with the problem? Do each of you have any ideas as to why this problem exists?" The purpose of such questions is to clarify the problem and to begin to obtain its behavioral history.

To place the information collected about specific problems into the context of their total sexual relationship, details are sought on the couple's entire "sexual script" (Gagnon, 1974). Key areas to investigate include:

1. Frequency of intercourse
2. Behaviors during foreplay
3. Duration
    Foreplay
    Time until orgasm/ejaculation
4. Sexual performance
    % of occasions orgastic
    Erectile difficulties
    Lubrication
    Ejaculation problems
    Pain
5. Positions during intercourse
6. Role of each partner
    Initiating sexual behavior
    During sexual behavior (e.g., active vs. passive)
7. Sexual setting
    Where
    When
    Variations

8. Behaviors after intercourse
9. Taboos or exclusions
10. Contraceptive practices
11. Idiosyncrasies

During the initial interview, the therapist tries to determine if general relationship problems contraindicate sex therapy. It should be kept in mind that many if not most couples who enter sex therapy have relationship or marital difficulties. These relationship problems may be due primarily to the stresses and strains placed upon the couple by their sexual problems. However, the relationship problems may be independent of the sexual concerns and thus may represent basic marital discord. Marital distress per se is thus not the sole criteria for deciding that sex therapy is inappropriate. Rather, sex therapy is contraindicated if the relationship is so unstable, or distressed that the couple will not be able to carry out the cooperative demands of the home assignment component of the sex therapy program (Leiblum & Kopel, 1977).

The therapist explains the general format and approach of the sex therapy program and answers the couple's questions at the conclusion of the intake session. If the therapist or the couple feels that the therapy program is inappropriate, alternatives are discussed. At this point, if any doubts regarding biomedical factors exist, the clients are instructed to consult a medical specialist (e.g., gynecologist or urologist) before proceeding. If the decision to continue is arrived at by all parties, the therapist schedules individual sexual learning history sessions for each partner and may give questionnaires to the couple to complete at home.

Sexual Learning History

A common feature to most behavioral sex therapy programs is the use of the sexual learning history interview (Annon, 1975a). Typically, this interview is conducted with each partner separately and lasts one to two hours. The interview is structured so as to cover major categories of sexual development, sexual practices, and attitudes. History-taking occurs within a behavioral framework; past experiences, attitudes, and emotions are seen as important as they relate to the current situation.

Many content outlines for the sex history interview are available (e.g., Hartman & Fithian, 1972; Masters & Johnson, 1970; Kaplan, 1974). We suggest the following outline as a guide for this specially designed interview:

1. Parents and home
   a. Relationship with parents and siblings
   b. Expression of affection
   c. Influence of religion
   d. Discussion about sex
   e. Overt and hidden messages about sex
   f. Nudity
2. Sexual development
   a. Sex education
   b. Earliest childhood sexual experiences
   c. Masturbation
   d. Menstruation, breast and genital development
   e. Fantasies and dreams
   f. Pornography
   g. Dating history—social relationships
   h. Influence of peers
   i. Intercourse (particularly first experiences)
   j. Homosexual experiences
   k. Unusual sexual experiences
   l. Traumatic experiences
   m. Situational factors
   n. Self-image, body image
   o. Premarital sexual experiences
3. Marriage
   a. Previous marriages and sexual experiences
   b. Courtship
   c. Early sexual experiences with current spouse
   d. Effect of children's births
   e. Current sexual attitudes
   f. Contraception
   g. Extramarital sexual behavior
   h. Physical attractiveness of spouse
   i. Relationship issues
   j. Romanticism
4. View of self
   a. Sexual competence
   b. Body image and acceptance
   c. Sexual guilts
   d. Expectations and goals
5. Miscellaneous
   a. Medical histories

b. Knowledge of spouse's sex history
   c. The individual's view of what caused the sex problem
6. Further investigation of areas from the sexual problem history and further examination of current sexual practices

Specific sexual problems are explored in detail. For example, for women who have never experienced orgasm, an assessment of specific fears regarding orgasm is often important. Common fears often relate to loss of control issues such as: (1) fear of being embarrassed; (2) fear of husband increasing his sexual demands once orgasm occurs; (3) fear of urinating while or just prior to the orgasm, actually based upon similarities between the internal antecedents of urination and orgasm for some women; (4) fear of being overwhelmed; and (5) fear of enjoying orgasm so much that "nymphomania" develops.

In cases where the sexual difficulty is either situation-specific, partner-specific, or previously not problematic, it is crucial to examine the details involved in both the success and failure occasions. For example, couples occasionally report that their chronic sexual problems did not occur on several occasions, such as during a vacation away from home without the children. Details of these experiences may suggest the importance of such factors as scheduling, distractions, daily pressures, relaxation, or a romantic context.

In one case, the discussions of partner-specific erectile failure clearly suggested that the major maintaining factor was a misattribution of erectile failure. Instead of viewing the loss of erection to the large amount of alcohol consumed, this individual thought the problem had some connection with his new partner. During their second sexual experience, he was totally distracted by fears of erectile loss and the implications for their new relationship.

During the individual interview sessions, the therapist may receive information that one client wants kept from the partner. Although there are drawbacks to permitting such "secrets," these concerns are balanced by the crucial information that may be provided. A not-uncommon example is that one partner admits to current or past extramarital affairs. Such information can be crucial if, for example, the "dysfunctional partner" is in fact sexually functional with a different partner. Thus, it is suggested that the individual interview session start with a statement from the therapist acknowledging both the awkwardness of getting such "confidential information" and the importance of receiving this information. The provision of confidentiality has led clients to confess

Table 6.2
Sexual Questionnaires and Inventories

Sexual Knowledge Inventory (McHugh, 1955, 1967, 1968)
Two forms available along with a counselor's manual for assessing knowledge and misinformation regarding human sexuality.

Sexual Response Profile (Pion, 1975)
Eighty items assessing sexual knowledge, attitudes, and current as well as past sexual behaviors.

Sexual Fear Inventories (Annon, 1975b, 1975c)
One-hundred and thirty items assessing extent of fear, discomfort, or displeasure for sexual behavior and experiences.

Sexual Pleasure Inventories (Annon, 1975d, 1975e)
Parallels format of Sexual Fear Inventory items assessing arousal or pleasant feelings.

Heterosexual Attitude Scale (Robinson & Annon, 1975a, 1975b)
Seventy-seven items assessing attitude or emotional feeling about sexual behavior and experiences carried out alone or with partner.

Heterosexual Behavior Inventories (Robinson & Annon, 1975c, 1975d)
Measures the range and frequency of sexual activity engaged in alone and with partner for the same 77 items as the Heterosexual Attitude Scale.

Sexual Interaction Inventory (LoPiccolo & Steger, 1974)
Composed of each partner's ratings on 17 specific sexual behaviors. Assesses heterosexual adjustment and satisfaction. Interpretations from a profile of 11 clinical scales.

to severe marital problems that the other partner hasn't acknowledged. In one case, the husband requested that some very unexpected information be kept confidential. He "confessed" that he had masturbated at the age of 14, but not since. Clearly, his selection of confidential information revealed much about his sexual attitudes and guilt.

Questionnaires and Inventories

A variety of questionnaires and inventories are available to assess sexual behaviors, attitudes, and knowledge. Table 6.2 lists a number of these assessment tools and gives brief summary descriptions. These paper-and-pencil devices offer the clinician an economical supplement to the interviews in collecting assessment information. The tests are objective and result in quantified scores which lend themselves to pretreatment-posttreatment evaluations. Furthermore, a review of responses to individual items permits the clinician to readily pinpoint deficits or excesses in sexual knowledge, attitudes, and behaviors which can be further explored in the interviews. The questionnaires can be given to the couple after the initial interview for them to independently complete at home. If the forms are mailed to the clinician, they can be reviewed before the individual sex-learning history sessions. Thus, the clinician can use the information from these forms for further inquiry or clarification during the individual history interviews.

The Sexual Interaction Inventory (SSI) developed by LoPiccolo and Steger (1974) is a particularly useful assessment device. The demonstrated methodological adequacy and clinical validity of this instrument distinguishes it from other sexual questionnaires and inventories. This inventory consists of 17 items covering the range of heterosexual behaviors from "the male seeing the female when she is nude" to "the male and female having intercourse with both of them having an orgasm (climax)." Each partner is instructed to independently respond to each item by answering six questions along dimensions of actual and desired frequency and pleasantness for themselves and their partner. The responses are scored and integrated into 11 scales to form a clinical profile for the couple. A sample profile with pretreatment data from a clinical case is presented in Figure 6.1. The Sexual Interaction Inventory can be used in several ways: (1) to assess dysfunction across multiple clinical scales; (2) to pinpoint specific areas and behaviors for treatment; and (3) to evaluate outcome results.[1] This instrument also has the advantage of assessing discrepancies between the partners in their attitudes and preferences regarding the sexual activities; thus, areas of misperception, insufficient communication, and basic disagreement are easily pinpointed and quantified by various scales.

The clinician may want to administer routinely the Marital Adjustment Scale (Locke & Wallace, 1959) or restrict its use to cases where marital discord appears to be a relevant factor. As described in Chapter 5, this test is very economical and it provides a good overall index of marital satisfaction by each partner. (See Chapter 5 for a copy of the Marital Adjustment Scale and the scoring key.)

[1] The Sexual Interaction Inventory, as well as a scoring and interpretation manual, is available from Dr. Joseph LoPiccolo, Department of Psychiatry and Behavioral Sciences, State University of New York Medical School, Stony Brook, New York 11790.

Figure 6.1 Sample Profile with Pretreatment Data (Sexual Interaction Inventory)

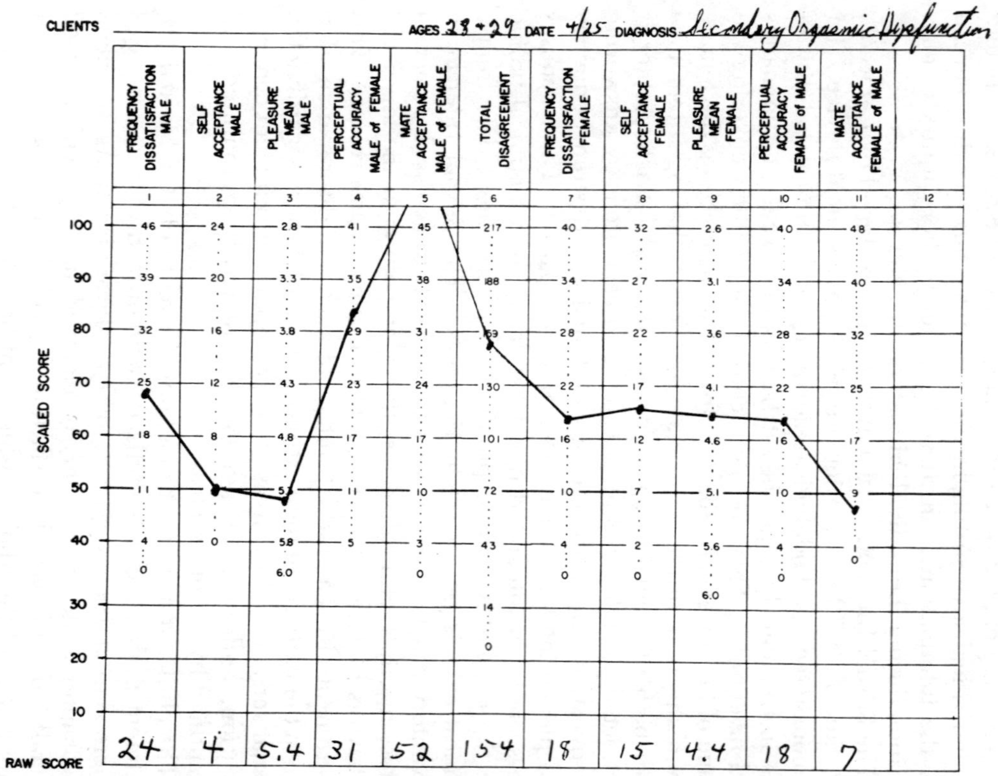

Reprinted by permission of Joseph LoPiccolo, Ph.D. Copyright Joseph LoPiccolo, Ph.D.

Measurement and Functional Analysis

In chapters covering other clinical problems, we review the use of naturalistic observations and laboratory analogue assessments as methods of measurement. For the area of sexual dysfunction, however, these procedures are usually inappropriate for a variety of obvious reasons. Rather, the primary sources of measurement are the behavioral interviews and paper-and-pencil instruments which focus upon recent behaviors and environmental factors.

When appropriate, a couple can be asked to collect baseline data via self-recordings before the start of the treatment program. For example, we have asked our clients for simple frequency counts of sexual sessions during the period between intake interview and start of treatment procedures. For cases of premature ejaculation, the female partner has been instructed to monitor and record the male partner's approximate time between entry and ejaculation for all occasions of coitus during the pretreatment period. The use of self-recordings, however, is limited to baseline data collection of behaviors which occur at reasonable frequencies. Unfortunately, couples seeking sex therapy often have current patterns of sexual behavior which are at or approach zero base-rates. In such situations, the use of self-recording is precluded and the therapist initially must rely upon estimated measurements from the interviews and questionnaires.

Once the sharing conference has taken place and the therapy contract explicitly has been negotiated, more direct behavioral measures can be collected by the couple during the treatment program. These measurement procedures will be discussed in a later section.

To prepare for the sharing conference, the clinician faces the complex task of integrating all the available information collected into a cohesive behavioral case formulation. This conceptualization is probably the most difficult and critical step in the behavioral assessment of sexual dysfunction.

Couples entering sex therapy are usually most concerned with the final response in the chain of physiological arousal (e.g., orgasm, ejaculation, or intercourse). For case formulation, it is important for the clinician to focus on events or conditions that are earlier links in that arousal chain. As a behavior, sexual arousal (and performance) is best understood as a chain analysis in terms of motor, cognitive, and physiological responses, as well situational factors. Furthermore, since sexual behavior is an interpersonal act, the factors of sexual communication skills and sexual arousal skills are clearly relevant. Table 6.3 presents a conceptual scheme to organize the assessment information in a way that will be helpful for planning treatment strategy. The clinician is reminded that categories listed in Table 6.3 are functionally interrelated. Using the S-O-R-K-C model, the clinician can conceptualize these interrelationships in a behavioral case formulation. To highlight this strategy, we have included some possibly relevant elements of the S-O-R-K-C formulation within parentheses adjacent to the categories listed in Table 6.3.

Matching Treatment to Client

By completion of the individual sexual learning histories, a wealth of information has been gathered. Although observation by others cannot be used, self-reports gain credence based on their consistency between partners. What Masters and Johnson (1970) have referred to as the "round-table discussion" is in a sense the closing of the first three stages of assessment. It is identical in purpose to the sharing conference described in other chapters. In this meeting, the therapist presents his conceptualization of the problem and discusses targets and overall treatment strategy. The goals of sex therapy are mutually agreed upon and specified by therapist and clients.

Often the goal stated by the couple is a specific final physiological response, such as orgasm. In these situations it is useful for the therapist to acknowledge that goal but to emphasize that earlier elements of the sequence need to be specified which maximize the likelihood that the final response will occur. Thus, subgoals are discussed, for example, in terms of feeling comfortable with the sexual situation, achieving high levels of arousal, and not being concerned about performance. These goals are broken down into specific target areas, such as communication skills and sexual arousal skills. A final goal that we recommend the therapist discuss with the couple is twofold: (1) that the couple learn how to identify sexual difficulties early, before they get out of hand; and (2) that the couple learn how to resolve or prevent these problems through self-interventions. The couple is told that, in essence, the final goal is to learn the information and skills necessary for them to preserve therapeutic gains over time without the future help of a sex therapist.

Typically, sex therapy for sexual dysfunction involves home

Table 6.3
A Framework for Integrating Assessment Information

1. Informational deficits (S-O)
   a. Sexual anatomy and physiology
   b. Base-rates: Norms and ranges of sexual behavior
   c. Misconceptions

2. Arousal skill deficits (R-K-C)
   a. Foreplay techniques
   b. Timing and coordination
   c. Positions of coitus
   d. Range of arousal techniques
   e. Active vs. passive roles
   f. Facilitators
   g. Blocks or inhibiting factors

3. Physiological performance and subjective pleasure (R)
   a. Erection—ejaculation
   b. Lubrication—orgasm
   c. Entry—pain
   d. Organic—medical factors

4. Perceptual problems (S-O-R)
   a. Distractions
   b. Fantasy ability
   c. Ability to sensate focus
   d. Observer of own behavior role

5. Relevant anxieties, fears, concerns, and guilt (S-R-K-C)
   a. Performance anxieties
   b. "Loss-of-control" fears
   c. Relationship concerns
   d. "Success" fears
   e. Areas of guilt or shame

6. Sexual communication skills (S-R-K-C)
   a. Ability to discuss sexual matters
   b. Ability to provide partner with specific positive feedback
   c. Ability to provide partner with *constructive* negative feedback

7. The sexual script (S-O-R-K-C)
   a. Current patterns of sexual activity
   b. Initiation of sex—By whom and how?
   c. Place(s) and setting(s)
   d. Props
   e. Events before
   f. Events after
   g. Competing activities

8. Avoidance/escape patterns (R)
   a. Frequency and duration of sexual activities
   b. Refusal

(cont.)

Table 6.3 (*cont.*)

    c. Resentments
    d. Positive and negative contingencies for escape/avoidance patterns

9. Body image and acceptance of own sexuality (S-O)
    a. Labels used for sexual self-concept
    b. Acceptance of body parts
    c. Views of sexuality
    d. Influence of religion and/or moral code
    e. Appropriate labeling of arousal

10. Expectations and goals for the sexual relationship (S-O)
    a. Ideals
    b. Realistic goals
    c. Motivation and commitment to program

11. The nonsexual relationship (K-C)
    a. Marital discord
    b. Relationship roles
    c. Implications for sexual satisfaction
    d. Implications for sex therapy

---

assignments which follow an in vivo, graduated approach. Thus, an important issue to discuss with the couple is the restriction of their sexual activity to the steps in the therapy program. The rationale of minimizing performances demands and building upon small positive steps of improvement, or *shaping*, is given. If the couple, however, cannot accept such limitations or structure, a more individually tailored treatment approach will be needed.

Annon (1974) has raised an important assessment and ethical issue regarding the use of a "package" total treatment program versus an individually tailored hierarchical treatment program for sexual dysfunction. Annon has questioned the validity of applying a total Masters-and-Johnson type sex therapy program to all cases of sexual dysfunction. Instead, he suggests applying his PLISSIT model in a hierarchical manner defined by sequential assessment and treatment of the following areas: (1) Permission; (2) Limited Information; (3) Specific Suggestions; and, if required, (4) Intensive Therapy. The major strategy in this model is to assess and apply intervention one step at a time, based upon the initial stages of assessment. We agree with the principle of this model, which encourages the therapist to use assessment information to decide on which components of an overall treatment program are relevant for a specific case.

The procedures available for treating sexual dysfunction have been clearly delineated (e.g., Masters & Johnson, 1970; Kaplan,

1974; Annon, 1974, 1975a). The task for the clinician is to conceptualize, set goals, and specify targets for change so that treatment procedures can be selected and individually tailored, based upon assessment information.

Assessment of Ongoing Therapy

In sex therapy, the assessment of ongoing therapy primarily involves collection of information concerning the couple's sexual behaviors during home assignments. The therapist is interested in whether and how the couple followed through on specific suggestions and in the effects of these experiences. Since direct naturalistic observations are not appropriate, the therapist usually relies upon interview reports by the couple during the office sessions.

We have found that the accuracy of these self-reports is often questionable, particularly when couples engage in more than one home sexual session per week. The partners may confuse the sessions and thus their recall of specifics is distorted. Couples have fallen into heated arguments over details when they disagree on what happened in a given session. To avoid such problems in monitoring ongoing assessment, we usually ask the couple to use self-recording procedures for each occasion of sexual activity.

The Daily Record Form (Lobitz & LoPiccolo, 1972) is one self-recording tool that we use to aid couples in reporting specific information. It is applicable to most home assignments that the couple will be given in sex therapy. Each partner is instructed to fill out the Daily Record Form as soon after a home session as is reasonable. We suggest that he or she include particular behaviors or experiences that were positive or negative. This self-recording form is useful in monitoring and measuring changes in sexual activity along with the changes in perceived pleasure and arousal. The form easily reveals meaningful changes or problems which are then explored further in the therapy session to pinpoint associated changes in behavior and attitude, and related environmental factors. The section for describing feelings and personal comments permits the client to express in writing important feelings that he or she might be unable to otherwise bring up during the therapy hour, e.g., feeling rejected by the partner or "turned off."

The Daily Record Form provides the therapist with information to assess whether the couple is ready to: (1) proceed to the next step of the therapy program; (2) remain at this step for another week; (3) return to a previous step; or (4) have specific modifications of the current steps.

Figure 6.2 provides an illustration of a couple's use of the Daily Record Form over two home sessions midway through treatment. The husband and wife both were in their late thirties and had been married for 12 years. They entered sex therapy with complaints of general sexual dissatisfaction, premature ejaculation for him and primary inorgasmic dysfunction for her. At the start of therapy, the couple had a very restricted sexual repertoire and a high level of inhibition as exemplified by the fact that she had never manually stimulated her husband's penis. As treatment progressed, the couple achieved their treatment goals of increased ejaculatory latency, female orgasmic ability, and general sexual satisfaction.

Self-recording procedures may also be used for assessing specific changes in the treatment of premature ejaculation. Couples are instructed in the squeeze technique (Masters & Johnson, 1970) and the female partner is asked to record the amount of time between the start of stimulation and the application of the squeeze for each trial. This latency data is recorded on a simple index card, noting the number of squeezes and each latency until the actual ejaculation. Reviewing these latencies permits the therapist to guide his treatment recommendations by the actual results for this particular couple. We usually graph the average latency per session to get an overall sense of progress. Couples have responded very positively to such graphs, commenting on the positive impact of this visual demonstration of improvement.

Observations of the couple role playing or practicing certain types of behaviors in the therapy office is another source of assessment information which may be useful for treatment. For example, after instructing couples on techniques of caressing, we usually request that they practice the touching on each other's hands and arms in the therapy office (cf. Hartman & Fithian, 1972). Thus, we assess whether they understand and can apply the principles of caressing and pleasuring. Additional instruction and practice is provided, if the observations suggest the need. Role playing is a particularly useful vehicle for the therapist's observation of sexual communication skills. For example, the couple is asked to simulate the situation in which one has verbally initiated a sexual session and the other partner refuses. The initiation-refusal issue is often a highly sensitive area for the couples. Observations of communications given by each partner will permit the therapist to assess actual or potential problems in this area. Such observations

Figure 6.2 Daily Record Form

NAME: _____ WEEK OF: May 30 TO June 8

Activity (specific behavior)	Date	1-10 Pleasure Rating	1-10 Arousal Rating	Description of experience feelings and personal comments (negative & positive)
Giver: I definitely initiated this session. I started by gently touching my husband's chest, neck, arms, and penis. He became aroused.	June 1	7	7	Giver: I thoroughly enjoyed what I was doing. I was aroused before we even got into bed.
Receiver: I enjoyed my husband's touching very much. I placed my hand over his during the genital stimulation. I tried to enjoy what was happening without worrying about reaching a climax.		8	8	Receiver: It was a beautiful session. We ended very close and contented. I did not have a climax but I wasn't as unhappy as I've been in the past. I did have an urge for my husband to insert his penis.

have helped us to pinpoint specific words, nonverbal cues, or styles of communication that led one partner to feel rejected or the other to feel guilty. The therapist can use information collected by these and additional observations to decide on and guide treatment procedures such as behavioral rehearsal.

Evaluation of Therapy

There are several assessment tools available for evaluating the outcome of sex therapy. First, the couple's self-reports of specific changes in their behavior, attitudes, feelings, and physiological responses can be compared to the pretreatment patterns and goals for these variables. For example, frequency and duration of intercourse are important variables to examine. The couple essentially determines what is a reasonable and desirable outcome.

For specific dysfunctions there are generally accepted criteria of success. In the case of erectile difficulties, the criteria of success is very likely tied to the ability to maintain an erection after entry long enough for mutual satisfaction. For cases of dyspareunia and vaginismus, a favorable outcome occurs when pain is no longer present or when entry is possible and mutual sexual satisfaction is achieved. For premature ejaculation, evaluation of therapy depends on a number of factors. Here, the critical issue for assessment is whether latency is sufficient for both partners to achieve mutual satisfaction, rather than the absolute number of minutes before ejaculation. Extremely short latencies, however, do suggest the potential for future difficulties. Latency to ejaculation is evaluated relative to foreplay activities and the timing of the female arousal responsiveness. If the female partner experiences difficulties in achieving orgasm after latencies to ejaculate are reasonably extended, the focus of therapy may shift to foreplay activities, coordination of entry between the couple or, in general, the enhancement of sexual arousal for the female.

Measures of outcome may also be obtained by readministering questionnaires or inventories. Pretreatment-posttreatment measures are invaluable to the therapist for research, clinical evaluation, and feedback to improve his or her own skills. As previously mentioned, we feel the SII is a particularly valuable paper-and-pencil instrument for such evaluations. The standardized scores permit an assessment of change for a particular couple relative to norms reported in the literature. Questionnaires are always readministered prior to termination so that areas requiring further clinical intervention can be identified and dealt with accordingly.

Perhaps the most difficult task of evaluation is to assess the prospects for maintenance of improvement. It has been our experience, as well as that of other sex therapy researchers (e.g., Lobitz & LoPiccolo, 1972), that therapeutic gains represent short-term changes if maintenance strategies are not part of the treatment plan. Lobitz and LoPiccolo (1972) suggest several procedures that are helpful in the assessment of stability of change in sexual functioning. First, as target goals are successfully met, the couple is asked to demonstrate their ability to deal with specific problems that could lead to "relapse." For example, if frequency has improved, the couple is asked to role play a situation in which one initiates sex and the other refuses for a variety of reasons. Thus, an assessment can be made regarding their ability to constructively cope with a potentially problematic situation. Second, the couple may be requested to write their own maintenance program at home over the course of several weeks. This exercise is broken down into several sections:

1. A list of "turn-ons" and "turn-offs" for self and partner
2. A list of danger signals (early signs of problems or returns to old patterns)

3. A summary of old and new patterns with a description of how old patterns led to or maintained problems and how new patterns preclude problems
4. A list of self-interventions (i.e., couple interventions)

Each part of the maintenance program is discussed in depth during subsequent therapy sessions. The completed maintenance program written by the couple, along with the couple's response to the progressive fading out of the structure of therapy, provides a direct assessment of the couple's self-control abilities (Kopel & Arkowitz, 1975).

Follow-up assessment is conducted in several ways. Masters and Johnson (1970) have used a telephone interview to assess long-term changes. However, such a procedure ignores the tendency for demand characteristics to distort the data. It seems that a telephone call or cover letter to explain the importance of candid answers to a written questionnaire is a better choice. A follow-up questionnaire should assess overall sexual satisfaction, specific target changes in sexual patterns and feelings, difficulties encountered since termination, and how the couple has handled or prevented any problems. It is suggested that the couple be given instructions that they will receive feedback on the follow-up questionnaire. The couple's motivation for filling out the questionnaire is probably enhanced if they believe the task to be part of the therapeutic process. In addition, if some responses are unclear or need further exploration, the couple can be told to expect future contact by the therapist. On occasion, the follow-up questionnaire has led by mutual agreement to booster therapy sessions for a couple.

In concluding, the reader is reminded that, in contrast to measurements for other clinical problems, assessment measures for sexual dysfunction are relatively crude. Aside from very limited in-office observation of role playing or practicing, the therapist must rely solely upon self-report data. The area of physiological laboratory assessment offers promising techniques of measurement (Rosen & Keefe, 1977) to supplement the self-report data. However, issues of reactivity, interpretation, availability of equipment, and privacy have not been dealt with adequately enough to suggest physiological strategies of assessment at this time.

CHAPTER 7

Behavioral Assessment of the Institutionalized Patient

Behavioral assessment of the institutionalized patient involves a characteristic set of methods and problems. In this chapter, we review methods of behavioral assessment particularly suited to institutionalized patients and explore the practical problems of conducting behavioral assessment in institutional settings. We discuss the methods and problems in the context of the five previously defined procedural stages of the assessment process.

Problem Identification

The behavioral problems of institutionalized patients are usually severe and complex in nature. Isolated, easy-to-define complaints are the exception rather than the rule. This makes problem identification and selection one of the most difficult, yet most essential, stages in the behavioral assessment of inpatients.

The behavioral clinician begins assessment by establishing general complaints. These complaints are then reduced to a list of discrete problematic behaviors. Using this list, behavioral goals for assessment and modification are chosen.

Interviews

Interviews with the patient and those who have contact with him provide a great deal of information regarding current problems and complaints. A major advantage of institutional settings is that multiple informants are available for interviews. Interviews are easily scheduled with the patient himself, attendants, other therapists, and members of the patient's family. Problems identified by one informant are substantiated or refuted by other informants.

Interviews follow the format described in previous chapters. Emphasis is placed on identifying current problems and complaints.

In institutional settings, informants typically have well-formed ideas and opinions as to the nature and causes of a patient's problems. Nurses, aides, and other mediators in the institution vary in background and training. Their past experience leads them to consider certain patient behaviors more problematic than others. In initial interviews with informants, the behavioral clinician tries to guard against the tendency to focus on certain problems to the exclusion of others. The main purpose of these interviews is to obtain a comprehensive description of the patient's behavioral assets and deficits.

Reviewing Patient Records

Psychiatric and medical records provide background information that is helpful in identifying and understanding current problems. Psychiatric records contain data pertinent to vocational background, marital status, family history, educational background, religious training, and sexual adjustment. Results of mental-status exams, psychological testing, and psychoneurological examinations are also typically included. Some notes on the patient's current functioning in the institution are also available. The utility of these notes for the behavioral clinician varies. Impressionistic notes are not very useful. However, notes based on actual observations of the patient may be very helpful.

The importance of medical records cannot be overstressed. The institutionalized patient often suffers from medical problems that underlie presenting problematic behaviors. Brain damage, infections, systemic diseases, and exposure to toxic substances have profound effects on behavior. Sensory-perceptual deficits may affect a patient's ability to respond appropriately to his environment. Maturational effects and physical disabilities place limits on the learning that can be expected to occur in behavioral treatment programs. The effects of the patient's medication regimen also need to be closely considered. Consultation with qualified medical personnel regarding medical problems is an important component in the behavioral assessment of institutionalized patients.

Questionnaires and Preliminary Observations

Questionnaires, checklists, and surveys are useful methods of establishing general complaints. The instruments reviewed in earlier chapters can be applied readily in institutional settings. The behavioral clinician may also develop referral forms for use within the institution. Referral forms such as those described by Alper and White (1971) ask for a brief description of problematic behaviors and the situational determinants of these behaviors. The forms are filled out prior to the initial interview and provide a helpful starting point for discussion.

If at all possible, the clinician also makes a brief preliminary observation of the patient. These observations are a rich source of data. Important aspects of the patient's environment that may have been omitted or overlooked by the patient himself, or by those working with him, are readily apparent when the clinician has the chance to observe the patient.

Problem Listing

Following preliminary data collection aimed at identifying general problems and complaints, a list of specific behavioral problems is constructed. In work with inpatients, behavioral clinicians often employ a *problem list* similar to that used in the Problem-Oriented Medical Record (Weed, 1968; Hayes-Roth, Longabaugh, & Ryback, 1972; Grant & Maletzky, 1972). The purpose of this list is to outline *all* of the behavioral problems that have been identified. Table 7.1 displays a format that can be used in making a problem list.

Each problem listed is assigned a number. Problems are defined in clear, measurable terms. Whenever possible, definitions are stated in the informant's own words. Problems are listed as they are identified. New problems are easily added and old problems deleted, as appropriate. A key feature of the problem list is the identification of informants who pinpointed the problem. Grant and Maletzky (1972) note the advantages of this procedure:

> We have found it important to differentiate those problems defined or agreed upon by the patient from those defined solely by his environment. For instance, "excessive religious talk," "difficulties following rules," and "unkempt personal appearance" were problems defined by the staff yet not agreed to by the patient. This differentiation becomes operationally important because, in our experience, patients have sustained greater effort towards solving those problems with which they themselves are in agreement. (p. 126)

Problems are categorized as behavioral excesses or deficits. Behavioral excesses are considered problematic because they occur either too frequently, too intensely, or over too long a time period. Behavioral deficits are considered problematic because they occur

Table 7.1
Behavioral Problem List

Problem Number	Description	Informant	Date Identified
Behavioral excesses			
1	high frequency of self-deprectory comments	ward attendant	9-17-75
3	talks about imaginary friends	ward attendant group therapist	9-23-75
4	overeats	patient patient's wife ward attendant	9-23-75
Behavioral deficits			
2	does not respond to greetings	ward attendant	9-17-75
5	unable to hold down steady job	patient patient's wife	9-23-75
6	sits alone in room during social activities	patient ward attendant	9-25-75

either too infrequently, are too low in intensity, or are too brief in duration.

Problem lists are usually typed and kept at the front of the patient's chart. They provide the clinician with a handy, comprehensive index to possible targets for treatment.

Measurement and Functional Analysis

In institutional settings, behavioral measurements are almost invariably gathered by means of observation. Observations are usually made by those who have daily contact with the patient, such as attendants, nurses, and other ward personnel. In some cases, members of the patient's peer group or family also function as observers. The behavioral clinician works closely with observers to develop measurement methods. The emphasis is on practical observation procedures that observers will accept and use.

General Considerations

Behavioral assessment of the institutionalized patient relies on frequent observations carried out over long time periods. The performance of observers is critical to this process. Training observers and maintaining their performance in the institution is a major task for the behavior clinician. Attending to some practical procedural details can make this task easier (Schaeffer & Martin, 1969).

First, observational materials are made readily available and easy to use for observers. Data sheets are printed up in advance. They are mounted on a heavy-duty surface, such as a clipboard. Clipboards are easy to write on and protect data from being lost, damaged, or destroyed. Clipboards are hung in a prominent position, for example, a nurses' station, between observations.

Procedural details are specified prior to observation. Details include the time of each observation, the setting, and any special observer responsibilities, such as reinforcing appropriate behavior if present. In order to ensure that these details are followed consistently by all observers, they are written down on the data sheet.

Prior to the start of data collection, the clinician demonstrates to observers how observation procedures work. This may be done while role playing or while actually observing a patient. After modeling the procedure, the clinician has observers follow suit. Pilot observations are very helpful. They clarify misunderstandings that may exist. They also provide a chance to make final corrections in procedure.

Once the actual observations have begun, the clinician makes daily checks with observers. This may consist of a brief visit to the observation setting or a short phone call. Daily checks are important. They give the clinician a chance to evaluate data as they are gathered. Procedural problems, such as absence of an observer or equipment failure, are easily identified and rectified. Daily data checks also serve to reinforce observers. Praise, expressions of interest, and special recognition to superiors can be used as rewards for a good job of data collection. Intermittent use of available material reinforcers, such as money, time off, or a written recommendation for job promotion, is also very helpful.

Methods of Observation

As might be expected, the observation strategies most widely used in the assessment of inpatients are those which are least costly and time-consuming for observers and also precise enough to yield reliable and valid data. Three methods of observation appear to fit these requirements: direct measurement of a permanent product, event recording, and time sampling.

Direct Measurement of a Permanent Product

Direct measurement is used in cases in which the patient's behavior lasts long enough to produce some sort of permanent, measurable record. For example, a direct measurement procedure could be used to record the number of pieces of work completed by a patient in a sheltered workshop program, the number of items of clothing left on the floor of the patient's room.

Event Recording

Many problematic behaviors of institutionalized patients do not last long enough to produce permanent records, for example, bizarre verbalizations, crying episodes, cursing, and temper tantrums. These behaviors must be observed as they are actually happening in order to be measured. Event recording is a practical method of accomplishing this task. In event recording, the observer simply counts the number of occurrences of the behavior during a given observation period.

Figure 7.1 displays data gathered on the frequency of "bizarre speech" responses of a 7-year-old schizophrenic (Browning & Stover, 1971). Staff in a residential behavior modification program took daily counts of the number of times the child spoke in a bizarre manner; each time, staff members would leave and record on a data sheet the tally and content of the boy's speech. As can be seen in Figure 7.1, this recording procedure inadvertently led to an extinction of the bizarre speech, presumably because it was ignored.

Event recording can also be used to measure several patient behaviors simultaneously. Agras (1976) used an event recording procedure with an 80-year-old patient to monitor: (1) unwarranted somatic complaints; (2) crying spells; and (3) excessive aspirin-taking. Staff members took counts of the three behaviors during their daily contacts with the patient. Data were gathered over the course of the patient's hospitalization. Crying episodes decreased in frequency early in treatment after the patient had been urged to talk more openly about the death of her husband. The staff observations suggested that this woman's somatic complaints were maintained by staff attention. When staff began ignoring somatic complaints, they readily extinguished. Aspirin-taking showed a

Figure 7.1 Example of Unstable Behavioral Baseline of Bizarre Speech in a Schizophrenic Child

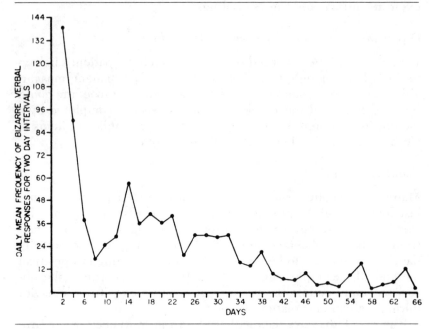

Reprinted by permission from Robert M. Browning and Donald O. Stover: *Behavior Modification in Child Treatment* (Aldine Publishing Co., Chicago). Copyright © 1971 by Robert M. Browning and Donald O. Stover.

similar reduction in frequency. Somatic complaints increased after a brief trial at home, but were reduced once again when the patient returned to the hospital environment.

Time Sampling

Time sampling is a third method commonly used for behavioral observations of institutionalized patients. In time sampling, a specified period of observation is divided into equal intervals. At the end of each interval, an observer, such as an aide or nurse, watches the target patient for a brief moment and records whether or not the behavior of interest occurred.

Time sampling is used to monitor behaviors which are ongoing or continuous. Since the observer monitors the patient for only a few seconds at a time, it is important that the definition used be clear and precise. Highly visible behaviors, such as head banging, rocking, walking, eating, and sitting are especially suited for measurement using a time-sampling procedure.

Time sampling is typically used in one of two ways: (1) to measure the percentage of time a patient engages in a single behavior, such as lying in bed, and (2) to simultaneously monitor multiple behaviors. This second approach has been extensively employed on behaviorally oriented inpatient wards.

The recording of multiple behaviors using time-sampling procedures has several advantages. First, it is an inexpensive method of data collection. Second, the recording of several behaviors simultaneously allows for comparisons to be made between operant levels of appropriate and inappropriate behaviors. Third, the procedure is comprehensive enough to be applicable to more than one patient. This minimizes the costs of training observers in a wide variety of behavioral observation strategies.

Several coding systems have been devised to categorize behavioral observations made using time-sampling procedures (Schaeffer & Martin, 1969; Harmatz, Mendelsohn, & Glassman, 1975; Liberman, DeRisi, King, Eckman, & Wood, 1974). The reader is urged to familarize himself with established coding systems before developing his own.

Matching Treatment to Client

Many different types of treatment are employed in institutional settings. Types of treatment used vary from institution to institution and may include individual psychotherapy, group therapy, psychopharmacotherapy, physical therapy, occupational therapy, or specific behavioral treatment programs. In the process of behavioral assessment, the clinician attempts to match treatment strategies to the needs of the patient. In work with inpatients, this process involves several important choices: (1) the choice of behavioral goals; (2) the choice of a drug treatment; and (3) the choice of a behavioral treatment program.

Behavioral Goal-Setting

Once problems are identified and measured, behavioral goals are set. Liberman and Bryan (in press) describe a method of goal-setting that is well suited for use in institutional settings. The clinician meets with the patient, his relatives, and mediators from the institution to review a list of presenting problems. Whenever feasible, problems targeted for modification are consistent with the eventual goal of facilitating the patient's reentry into community settings. Goals thus may be educational, vocational, or social in nature. Emphasis is placed on the selection of short-term goals that can be attained in a matter of weeks or months. Goals are arranged over time according to their difficulty. For example, during the early weeks of therapy, goals that are minimally difficult are set. In later weeks, progressively more difficult goals are set. The institutionalized patient often has a long history of failure experiences. In goal-setting, small increments of behavior change are required over short time periods. Success experienced by the patient when these goals are attained is an important motivational factor.

Meeting with multiple informants to establish goals does pose some problems. In most cases, the goals identified by the patient are given primary importance (Bandura, 1969). In some cases, however, patients are confused because of severe deficits in reality-oriented behavior. Patients may be unable to effectively communicate or select appropriate goals. Two approaches are helpful in dealing with this problem. The first involves working closely with the patient to specify easy-to-understand, concrete goals. Fairweather, Sanders, Maynard, and Cressler (1969) found that chronic schizophrenics can participate in goal-setting if presented with a variety of goals stated in terms that are simple and pertinent to the patient's daily living routine. The second approach is to work closely with the patient's legal guardians in setting goals for therapy.

Determining the appropriateness of stated goals is another problem. Behavioral clinicians need to be wary of selecting goals that are important to institutional mediators but that may have little relevance to the patient, e.g., quiet, docile, "good" patient behavior. Ayllon and Azrin (1968) have suggested that behavioral clinicians who work in institutions use the Relevance-of-Behavior Rule: *Teach only those behaviors that will continue to be reinforced after training.* Appropriate behavioral goals focus on responses which, if developed, have some likelihood of being maintained in current and target environments. For example, if the patient gains some proficiency in communication skills he is likely to be reinforced for this not only in the immediate ward environment, but also in community settings.

Choosing a Drug Treatment Program

Psychotropic drugs are an important component of treatment programs of institutionalized patients. Drugs are effective agents in dealing with many clinical disorders. For example, they can reduce anxiety, help patients cope with acute depression, and aid in the management of schizophrenic episodes. The use of medication as part of a comprehensive behavioral treatment program has been discussed in the literature (Fay, 1976; Liberman & Davis, 1975). Fay (1976) provides a useful list of questions that need to be considered by the clinician in implementing any drug program:

1. What is the nature of the problem?
2. Is the difficulty acute (lasting a few hours, days, or weeks) or chronic (continuous over a period of years)?
3. If acute, has the problem occurred on several previous occasions?
4. Which drugs have been taken by the patient before and with what results?
5. In what dose has the drug been used before?
6. Is insomnia a prominent feature of the symptomatology?
7. Is there a history of drug abuse?
8. Are strong suicidal tendencies present?
9. Is there a history of noncompliance with prescribed medication regimens?
10. What is the age of the patient and the general level of health, especially with regard to heart, blood pressure, kidney, and liver, as well as the presence of prostate enlargement, glaucoma, or drug allergies? (pp. 74–78)

Answers to these questions help the clinician tailor drug treatment programs to the needs of the patient. Behavioral assessment is then used to analyze the effects of drugs on behavior (Liberman, Davis, Moon, & Moore, 1973; McPherson & LeGassicke, 1965).

The clinician involved in behavioral assessment in institutional settings needs to be familiar with the principles and issues of drug treatment.

Choosing a Behavioral Treatment Program

In choosing an intervention strategy for modifying behavior in an institutional setting, several factors need to be considered. These include: (1) the nature of the target behavior; (2) legal issues; (3) the resources of the patient; (4) the resources of the clinician; and (5) the resources of the institutional setting.

The Nature of the Target Behavior

The nature of the target behavior dictates, to a certain degree, the specific treatment strategies that are employed. If the patient does not have the appropriate target behaviors in his repertoire, for example, vocational skills, intervention techniques designed to facilitate the learning of new skills, such as shaping and positive reinforcement, are indicated. If the patient's behavior is problematic because it occurs too frequently (hand washing), too intensely (screaming), or over too long a time period (excessive sleeping), techniques aimed at decelerating behavior, such as extinction or punishment, are indicated. When patients engage in behaviors that are so severe that they are life-threatening, treatment procedures are selected that work quickly to stop the behavior—for example, aversive conditioning.

Legal Issues

Legal considerations also play a role in the selection of intervention strategies. When designing a treatment program for a voluntarily institutionalized patient, the most important legal consideration is obtaining informed consent. The basic elements of informed consent (DHEW, 1971) are:

1. A fair explanation of the procedure to be followed, including an identification of those which are experimental;
2. A description of the attendant discomforts and risks;
3. A description of the benefits to be expected;
4. A disclosure of appropriate alternative procedures that would be advantageous for the subject;
5. An offer to answer any inquiries concerning the procedures;
6. An instruction that the subject is free to withdraw his consent and to discontinue participation in the project or activity at any time. (p. 7)

When working with involuntarily institutionalized patients, the legal issues are more complex. This complexity is illustrated in the decision made in the case of *Wyatt* v. *Stickney* (Wexler, 1973). The court's decision in this case places restrictions on many of the basic techniques that can be used in therapy. Table 7.2 reveals the scope of this ruling.

None of the rights enumerated under the section on reinforcers can be made contingent for involuntary patients. The punishments listed are under similar restrictions. Target behaviors for such involuntary patients can no longer include tasks related to the maintenance of the institution. Similar rulings have placed restrictions on the behavioral treatment techniques that can be used with institutionalized juveniles and violent prisoners.

State and federal mental health agencies are concerned about legal rulings regarding the use of behavioral treatment techniques

Table 7.2
Elements of the Behavioral Approach Affected by *Wyatt v. Stickney* Decision

Element	Decision
Reinforcers	
1. Primary reinforcers	Patients must receive, at the minimum, a diet meeting the Recommended Daily Dietary Allowance
2. Activity reinforcers	The opportunity for religious worship shall be accorded to each resident who desires worship
	Residents shall have a right to regular physical exercise several times a week
	Residents shall have a right to be outdoors daily
3. Material reinforcers	Each resident shall have an adequate allowance of his own clothing, or clothing supplied by the institution
	Residents shall sleep in single rooms or rooms of no more than six persons; screens or curtains shall be provided to ensure privacy. Each resident shall be furnished with a comfortable bed, a closet or locker, and appropriate furniture
4. Social reinforcers	Patients have the right to telephone communication and to send and receive mail
	Patients shall be given suitable opportunities for interactions with members of the opposite sex
Punishments	
1. Corporal punishment	Corporal punishment shall not be permitted
2. Restraint	Phsyical restraint shall not be employed as a punishment
3. Seclusion, time out	Seclusion, defined as the placement of a resident alone in a locked room, shall not be employed. Legitimate "time out" procedures may be used if systematically applied in a behavior-shaping program under direct professional supervision

(cont.)

Table 7.2 (cont.)

Target behaviors

1. Jobs within the institution	No resident shall be required to perform labor which involves the operation and maintenance of the institution. Residents may engage in such labor voluntarily if they are compensated in accordance with the minimum wage laws
	No resident shall be subjected to a behavior modification program which attempts to extinguish socially appropriate behavior or to develop new behavior patterns when such behavior modifications serve only institutional convenience

Therapy procedures

1. Physician approval	No resident shall be subjected to a behavior modification program without prior certification by a physician that he has examined the resident and finds that such behavior is not caused by a physical condition which could be corrected by appropriate medical procedures
2. Individualized treatment	Patients have the right to individualized treatment programs. These programs should be formulated by the institution and will include: a statement of specific needs and limitations of the patient, long- and intermediate-range goals; a statement of the least restrictive alternative for treatment with a projected timetable for their attainment. Programs will be continuously monitored using objective indicators. Each resident shall have an individualized post-hospitalization plan.
3. Institutional review of therapy programs having aversive elements	Behavior modification programs involving the use of noxious or aversive stimuli shall be reviewed and approved by the institution's Human Rights Committee and shall be conducted only with the express and informed consent of the affected resident, if the resident is able to give such consent, and of his guardian or next of kin, after opportunities for consultation with independent specialists and with legal counsel.

in institutions. Many of these agencies have developed directives intended to safeguard the rights of involuntarily institutionalized patients. These directives match, and in some cases, go beyond those outlined in *Wyatt* v. *Stickney*.

The practicing clinician needs to keep abreast of legal developments relevant to the practice of behavior therapy in institutions. We highly recommend a review of recent articles (Begelman, 1975; Wexler, 1973) and books (Martin, 1975) in this area. In certain cases, the clinician may need to consult with legal counsel or with representatives from committees on legal and ethical issues of professional organizations such as the American Psychological Association, the American Psychiatric Association, and the Association for the Advancement of Behavior Therapy.

The Resources of the Patient

The ability of the institutionalized patient to actively participate in treatment programs is a major factor in the selection of therapy techniques. Active involvement in therapy depends on factors such as age, intelligence, physical condition, and motivation. When working with patients who are young, fairly intelligent, and motivated, an emphasis is placed on self-control procedures such as relaxation, desensitization, or cognitive restructuring (Lazarus, 1971). When working with patients, such as chronic schizophrenics, who have severe deficits in reality-oriented behavior, intervention procedures that rely on environmental control, for example, token economies, are much more likely to succeed.

The Resources of the Clinician

The choice of treatment is also influenced by the practitioner's background and experience. If the clinician has limited training in behavior therapy, his repertoire and familiarity with certain treatment strategies also is likely to be limited. With therapists who have more training and experience, available treatment alternatives are expanded. The clinician whose experience is limited should start "small" with relatively simple treatment techniques, such as the operant reinforcement of appropriate speech. The move to progressively more difficult treatment strategies, for example, a ward-wide token economy, can be made as experience is gained. When working in an institutional setting, the clinician should remember that nothing succeeds like success. New and different treatment procedures should be introduced slowly so that success is experienced at nearly every step.

The practicing clinician also needs to consider his own time and materials in choosing a treatment strategy. Actually carrying out or supervising behavioral treatment programs in an institutional setting requires a considerable investment of time and energy. The clinician needs to remember that behavioral principles apply to him as well as to the patient. If the clinician's efforts are to be maintained, they must be reinforced and supported by the institutional environment. Material and financial support is of fundamental importance.

The behavioral clinician attempts to realistically appraise how his own resources affect options for treatment. His first, and perhaps most important commandment in this regard is: "KNOW THY OWN CONTROLLING VARIABLES" (M.J. Mahoney, 1974).

Resources of the Institution

The material and social resources of the institution are important determinants in the selection of a treatment program. The physical design of rooms and living areas, the arrangement of furniture and decorations, and the number and kind of recreational or vocational materials need to be considered in implementing treatment strategies. The resources of the various social groups that make up the institution also need to be considered. If staff members are trained in behavior therapy and motivated, the choice of treatment options is nearly unlimited. If staff members lack experience but are open to behavioral modification procedures, training programs may be implemented.

A major fact of the behavioral clinician's life is that he needs to rely on others within the institution. Cautela and Upper (1975) note:

> In our experience, it is futile to treat a client in an institution without ward cooperation in terms of record-keeping, continuing treatment procedures employed in private sessions, and consequating appropriate and inappropriate behavior in a systematic manner. (p. 288)

Those experienced in the practice of behavior therapy in institutions (Atthowe, 1973a, 1973b; Browning & Stover, 1971) also underscore the importance of coordinating the social resources of the institution. There is a clear need to shape institutional systems to accept and support behavioral programs. Consideration of this process is a major factor responsible for the "success" or "failure" of treatment programs such as token economies (Hall & Baker, 1973).

Assessment of Ongoing Therapy

Once an intervention procedure is selected and implemented, behavioral measurements are used to evaluate its effectiveness. Measurement procedures are identical to those used in earlier stages of assessment—for example, event recording or time sampling. Data from these earlier measurements provide a baseline record of the strength of behavior. Measurements of behavior taken during therapy intervention permit comparisons with baseline to be made. Measurements are typically taken from the moment a treatment program starts and are continued throughout treatment; they can be repeated periodically even after treatment has stopped, for purposes of follow-up.

If the strength of behavior shows a reliable change once a treatment program has begun, there is reason to believe that the intervention used was effective. For example, if contingent praise for appropriate speech is associated with a major reduction in bizarre verbalizations, one tentatively may conclude that this intervention is helpful. For clinical purposes, a comparison of baseline vs. intervention effects is often sufficient. However, this type of evaluation does not rule out the possibility that behavior change is due to extraneous influences not under the clinician's control.

Experimental analysis permits a more precise evaluation of therapy procedures. Within-subject designs, such as multiple-baseline or reversal designs, can be employed for the intensive analysis of the single case. The application of these designs is discussed in earlier chapters and also in the research literature (Sidman, 1960; Barlow & Hersen, 1973; Risley & Wolf, 1972). Within-subject designs are well suited to the analysis of behavior in environments which can be precisely controlled. The appropriateness of these designs in field settings where total environmental control is not feasible is questionable. Liberman, King, & DeRisi (1976) point out several problems with these designs:

1. Single-case designs require many repeated measurements to be made over time. Experimental analysis is incomplete if time runs out—for example, if the patient is discharged or transferred to another ward.

2. In order to maximize the power of single-case designs, stable baselines are needed prior to intervention. Stable baselines are difficult to achieve. Patients often enter institutions in a crisis stage and their behavior changes quickly in response to aspects of the treatment milieu that are beyond the clinician's control.
3. Ethical and legal restrictions make it difficult, if not impossible, for clinicians to exercise the degree of control over patients and mediators needed for precise experimental analysis. Thus, while sophisticated within-subject designs provide the most rigorous evaluation of ongoing therapy, it may be impractical to implement them.

Behavioral clinicians are firmly committed to taking behavioral measurements during therapy. These measurements form the basis for evaluating therapeutic effectiveness. The conclusions that can be drawn depend on the control over behavior that is demonstrated. Tentative conclusions can be drawn when reliable changes are evident in comparing data from a single baseline and intervention period. Stronger conclusions can be made if interventions are systematically manipulated over time—for example, instituted, stopped, and then reinstituted. When conducting assessment in institutions, the necessity for experimental confirmation of therapy effectiveness always should be weighed against practical considerations such as time, cost, and feasibility.

Evaluation of Therapy

The final evaluation of therapy is one of the most important stages in the behavioral assessment of institutionalized patients. This comprehensive evaluation is typically conducted just prior to termination with the patient. An evaluation interview session is scheduled with the patient and those who have been involved in his treatment. The topics reviewed include: (1) the outcome of therapy; (2) maintenance of behavior changes; and (3) generalization.

The outcome of the treatment program is easily determined by referring to the behavioral measurements made over the course of assessment. Obtained improvement is compared to desired improvement, and recommendations are made regarding the appropriateness of continuing versus terminating.

Maintenance of behavior change is critically important when working with institutionalized patients. If target behaviors have been chosen using the Relevance of Behavior Rule (Ayllon & Azrin, 1968), whatever new behaviors have been learned should currently be reinforced by the treatment environment. If this is not the case, a reevaluation is indicated.

Generalization of therapy effects is another crucial consideration in evaluating therapy. Generalization refers to the degree to which changes in target behaviors are mirrored by changes in other behaviors and in other settings. When working with institutionalized patients, generalization is often hard to achieve. A patient may learn a whole range of self-care behaviors during therapy, such as grooming, washing, or shaving, but show little change in his ability to interact appropriately with others. The degree of generalization obtained depends greatly on the resources of the patient and the therapy procedures used. If the patient has few behavioral strengths and the therapy relied heavily on environmental control procedures, such as a token system, generalization to other settings probably will be poor. If the patient took a more active role in therapy, learned behavior is more likely to be under self-control rather than environmental control; generalization to other settings is more likely to occur. When working with institutionalized patients, failure of behavioral effects to generalize is not uncommon. When generalization fails to occur, further work on other target behaviors and in other treatment settings is indicated. For example, mentally retarded patients who show progress in a behavioral treatment program in an institution may need to undergo similar treatment experiences in noninstitutional settings, such as a community-based sheltered workshop, halfway house, or social club.

The eventual goal of behavioral assessment and therapy is to return the patient to the community. In most cases, therapy is not considered completed until this goal is reached. The behavioral clinician needs to be familiar with the broader environmental setting of the institution. The availability of treatment settings in the community and the behavioral requirements of these settings are important considerations. The behavioral assessment of inpatients need not be and should not be restricted in focus only to the institutional setting (Fairweather 1967; Fairweather et al., 1969; Mischel, 1977).

PART III

Current Perspectives and Future Directions

CHAPTER 8

Behavioral Assessment of Systems and Society

Behavioral assessment has long been applied to individuals who exhibit deviant or undesirable behavior. Skinner (1971) has maintained that this approach can be broadened in scope to deal with larger social problems. A small but growing number of behavioral scientists are involved in efforts to "stretch" the behavioral model of assessment to fit the demands of community and societal applications.

Chapter 8 begins with a review of applications of behavioral assessment to industrial systems, educational systems, living environments, community problems, and social and legal reforms. The review is followed by a discussion of practical, theoretical, and ethical problems inherent in behavioral assessment at the societal and systems level.

Broad Applications of Behavioral Assessment

Assessment in Industrial Systems

Early attempts to apply behavioral assessment principles to industrial systems were limited. Demonstration projects were conducted with individual subjects (Verhave, 1966; Cummins, 1966) and there was speculation about fruitful areas for investigation (Aldis, 1966). Behavioral scientists have moved well beyond this stage. They are now grappling with more pervasive behavioral problems that occur in industrial systems.

Unemployment

A problem for many workers these days is finding a job. In conjunction with a state employment service, Jones and Azrin (1973) evaluated the effectiveness of two types of newspaper advertisements in landing jobs for unemployed workers. One advertisement asked people in the community to call the employment service if they knew of an available job. A second advertisement offered community informants $100 if they provided information that led to a job for an unemployed worker. The effects of the ads on the behavior of community residents were assessed using a simple procedure. Job counselors at the employment service kept records on the number of phone calls received from community informants, the number of job openings reported, and the number of unemployed workers actually hired. The advertisement offering a reward to informants produced significantly more jobs than the advertisement offering no reward. This study demonstrates the utility of simple record-keeping procedures in analyzing the effects of an intervention aimed at a large geographical area.

Tardiness

A problem plaguing employers in many industrial settings is chronic worker tardiness. A recent study by Herman, deMontes, Dominguez, Montes, and Hopkins (1973) examined this problem in a large factory. Worker tardiness was measured using the factory time clock and punch cards. The time clock is an excellent assessment device. It allows for precise, automatic recording of behavior. Tardiness in one group of chronically tardy employees included in the study dropped when a monetary bonus for punctuality was provided. Tardiness increased in a second no-treatment control group over the same time period. Similar incentive systems have been found to reduce worker tardiness in settings ranging from a hardware company (Nord, 1969) to a state legislature (*Newsweek*, July 7, 1975, p. 21).

While incentive systems may work, several issues need to be considered when they are used on a large scale. First, maintenance of punctuality once monetary incentives are withdrawn needs to be thoroughly evaluated. In order to be cost-effective, the behavioral effects of such bonuses should be long-lasting. Second, one needs to consider the effects that giving bonuses to chronically tardy workers may have on other employees. Incentive systems that limit their focus to the tardy may have unintended effects—workers who rarely have been tardy may become tardy to gain access to an incentive program.

Job Performance

Even if workers arrive at work on time, they may not begin their assigned work or perform at 100% efficiency. Several methods have been used to study job performance behaviorally. Pierce and Risley (1974) employed direct observation. Adolescents working as recreation aides in a Neighborhood Youth Corps program served as subjects. Trained observers used a daily checklist to tally whether workers were present in assigned work areas and had completed assigned jobs. Daily checks of worker performance were made under several conditions: (1) explicit statement of job description in terms of time and task requirements; (2) threat to fire if work did not improve; and (3) payment based on amount of time work completed. Payment for work completed resulted in nearly 100% completion of assigned tasks, whereas payment based on clock time produced the lowest level of work for any of the conditions, 35% completion of assigned tasks.

Automatic recording devices provide a second measure of job performance. Electronic transducers may be placed in tools to monitor their use. Appropriate use of such "wired" tools completes an electronic circuit and triggers other recording devices such as counters or timers. In this way, data are easily gathered on the number of times a tool was used per day or the amount of time it was used. Automatic recording devices are especially helpful if tools are used repeatedly to complete tasks, for example, as on an automobile assembly line. Tools that automatically record job performance have been used in a sheltered workshop (Schroeder, 1972) and in individual work with retardates (Tate, 1968).

In many cases it is not feasible or desirable to directly sample workers' behavior. Workers are often uncomfortable about being directly observed on their jobs and may refuse to cooperate. Skilled or semiskilled workers are likely to resent having to work with tools that automatically record their performance. Labor unions provide organized resistance to the widespread application of such measures. When faced with such realistic problems, other measures may provide the only means to assess job performance behaviorally. For some jobs, permanent product measures are easy to obtain. For example, it is easy to record the number of items tagged by a ticketer in a clothing factory, the number of cars coming off an assembly line in an hour, or the number of pounds of mail delivered by a mailman. Measures like these are gathered routinely in industrial settings.

While the assessment of industrial systems is a relatively new area for the behaviorally oriented practitioner, social scientists have been active in this field for many years. There is much to be learned from industrial psychologists and sociologists regarding basic phenomena and methods of research. Applied behavioral analysis may be a promising approach to assessment in industrial systems. A great deal more research is needed before this promise is fulfilled.

Assessment in Educational Systems

The principles and technology of behavioral assessment have been applied to educational settings for some time. Historically, the focus has been on one part of the educational system—the classroom. More recently, behavioral assessment has been applied to pervasive problems that affect the broader educational system.

School Discipline

The school principal is typically responsible for maintaining student discipline. As a result, he is often placed in a position in which negative behaviors are the primary focus in interactions with students. Working in a crowded, inner-city school system, Copeland, Brown, and Hall (1974) examined the results obtained when principals used a more positive approach. The emphasis in this approach was on praise and attention for appropriate behavior, rather than punishment for inappropriate behavior. In one study, a principal attempted to alter rates of attendence in chronically absent children. Data on student attendance were gathered as part of the school's daily routine. During baseline, target students attended school 41%–53% of the time. During the intervention phase, the principal visited the classrooms of target students and praised appropriate behavior. Daily praise took less than 60 seconds per child. Under the principal-praise condition, attendance increased to 79%–85%. In a second study, the performance of three low-achieving students improved when they were sent to the office to be praised by the principal for academic achievements. In a final study, the effects of principal praise on academic work of students in two third-grade classes were evaluated. Performance on daily arithmetic tests improved dramatically when the principal made visits to the classrooms to praise students' performance. In each of these studies, experimental control was demonstrated using a multiple-baseline design.

Misuse of Educational Materials

Students often use educational materials in an inappropriate fashion that deprives others of the opportunity to use them. School vandalism and the loss or wanton destruction of books and other materials are examples of this behavior. The misuse of educational materials is particularly a problem in libraries.

Meyers, Nathan, and Kopel (1977) focused on one aspect of this problem—the reshelving of library journals. The setting for this study—a university library—is one in which access to recent journals is very important. The problem of unshelved journals inconvenienced both library users and staff. For purposes of assessment, daily counts of the number of bound journals not reshelved were made by library staff. Figure 8.1 demonstrates that during baseline conditions and during an instructions condition in which signs reading "Please Reshelve Journals" were posted, the number of journals reshelved was rather low. When a token system was instituted, the number of unshelved journals dropped dramatically. The token system was a self-regulated one. Library users gave themselves a token for each journal reshelved. Tokens could be refunded for food, movies, photocopying services, etc. The costs of this program were low; the benefits gained were high. Assessment methods took little staff time and were sensitive to behavior change. Intervention procedures were so effective that staff members who normally spent time reshelving journals were free to work in other areas of the library. The library reinstituted the token program after the study had been completed, and a one-year follow-up reveals that progress has been maintained. In this case, an experimental evaluation led to permanent adoption of an effective program.

Racial Integration

Racial integration is a problem facing many school systems in this country. Mandatory school busing leads to physical integration of children of different races, but often fails to produce the desired social integration. Fostering interracial social interactions is thus a practical problem for many educational systems. Hauserman, Waley, and Behling (1973) explored the possibilities of a behavioral approach to this problem. Daily records of interactions between children of different races were made during a lunch period by a first-grade teacher. Baseline records indicated a very low

Figure 8.1 Daily Count of Unreshelved Journals Corrected for Library Use

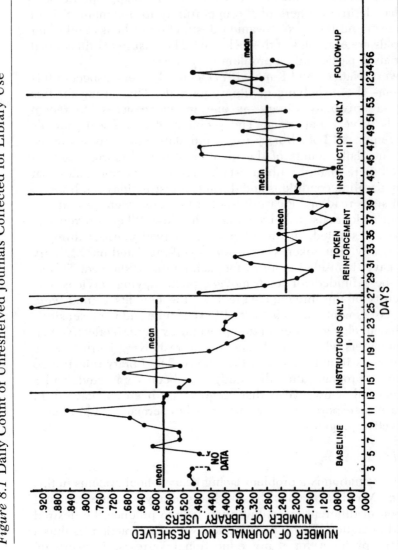

Reprinted by permission from Meyers, H., Nathan, P., and Kopel, S. Effects of a token reinforcement system on journal reshelving. *Journal of Applied Behavioral Analysis*, 1977, *10* (2), 216.

level of interracial interactions. During intervention, teachers prompted students to sit with a "new friend" opposite in race and reinforced the occurrence of this behavior. The prompting plus reinforcement effectively increased the level of interracial social interactions in the lunchroom. Equally significant was the finding that the effects of the lunchroom intervention generalized to a free-play period. That is, more interracial play occurred during free time even though students were not prompted or reinforced for this behavior. A major strength of this study was that interracial interactions were monitored across several different settings. This procedure is especially useful in assessing complex social systems in which interventions introduced in one part of the system may exert effects in other parts of the system.

Assessment of Living Environments

The behavioral approach to environmental assessment typically follows several steps. First, some aspect of the environment is selected for study, such as architecture, organization of furniture, availability of materials, schedules of activity, or patterns of staff assignment. Second, measurements of behavior are conducted under current environmental conditions. Third, the behavioral effects of systematic changes in the selected environmental variable are assessed. Finally, the information gathered may be used to make decisions about the design of the living environment studied.

In the past few years, a number of multidisciplinary research teams have developed behavioral measures for the evaluation of living environments. The Living Environments Group at the University of Kansas headed by Todd Risley has actively explored three measurement categories: activity measures, interaction measures, and stimulation measures. Together, the measures comprise the Resident Activity Manifest (Cataldo & Risley, 1974). The measures are designed to describe behavior in a wide range of environments.

Activity Measures

Activity measures assess the degree to which individuals participate in planned activities in their environment. The Planned Activity Check (PLA-CHECK) is a good example. The PLA-CHECK is a time-sampling observation procedure that is easy to use. First, a list of appropriate planned activities for a given environmental area is drawn up. The activities considered appropriate differ, depending on characteristics of the environment and subject's age, sex, and physical condition. Second, at given intervals, perhaps every 15 or 30 minutes, an observer walks through the designated area and makes two counts. One count records the number of individuals engaged in planned activities. A second count records the total number of individuals present in the area. A percentage participation measure is obtained for each observation. The percentage is calculated by dividing the number of individuals engaged in planned activities by the total number of individuals in the area and multiplying the result by 100. The percentage participation measure is an index of resident activity.

LeLaurin and Risley (1972) used the PLA-CHECK method to systematically evaluate the effects of different staffing patterns on the amount of time children engaged in teacher-planned activities. Data were collected in a day care center under two conditions: a "man to man" staffing pattern in which teachers kept children together in one activity until all had finished and a "zone" staffing pattern in which teachers sent a child on to a second teacher as soon as the child had finished a given activity. Participation in planned activities was much higher when a "zone" staffing pattern was used.

Activity measures also provide a means of evaluating the effects of architectural changes. For example, Twardosz, Cataldo, and Risley (1974) have used activity measures to assess the effects of open vs. partitioned rooms and noisy, light areas vs. quiet, dark areas on the behavior of children in a day care center.

Interaction Measures

Interaction measures provide data on what individuals are doing. The interaction measures described by Cataldo and Risley (1974) are most useful in environments in which participation in planned activities is low. Interaction measures are used to determine the amount of time individuals actually interact with materials, e.g., play with, manipulate, or repair, and to contrast this with the amount of time spent in simple contact with materials, e.g., holding or touching them. Continuous recording systems can be used to gather reliable data on interaction with materials (Cataldo & Risley, 1974).

Interaction measures can be used to determine which environmental factors influence behavior. For example, in one study reported by Cataldo and Risley (1974), data were collected on the behavior of retarded adults in a nursing home. Observations were conducted in a lounge area in which recreational materials were freely available. Under baseline conditions, the percentage of time spent interacting with materials was very low and the time spent in contact with materials was only slightly higher. During the intervention phase, a music activity was introduced. Volunteers played musical instruments, sang, and passed out instruments to residents. The music activity produced a marked increase in both contact and interaction with materials.

Stimulation Measures

Stimulation measures yield data on what individuals in a given environment are experiencing. This measure is most useful in describing behavior in environments where resident activity is extremely limited. For example, Cataldo and Risley (1974) employed a stimulation measure in a geriatric nursing home. Data were gathered on each nursing home resident hourly from 7 A.M. to 7 P.M. To measure stimulation, observers recorded the patient's location in the home (private room, hall, lounge, dining area, outdoors, geriatric chair area), his position/motion (sitting, wheeling, etc.), and speech. Residents in the nursing home spent most of their time in their rooms, sitting or lying down, and speaking with no one. Activity measures confirmed the results gathered by means of stimulation measures. Only 13% of the residents were found to engage in any sort of activity, even though activity was broadly defined to include any type of appropriate involvement with equipment, materials, or other persons.

Results from the stimulation and activity measures strongly suggested the need for changes in the nursing home environment. Redesign of this living area produced major changes in the patterns of resident activity. In this instance, stimulation measures served as the first step in a more comprehensive behavioral assessment.

Activity measures, interaction measures, and stimulation measures are practical procedures useful in describing the behavioral attributes of diverse environments. Similar measures have been used to assess the behavioral impact of environmental manipulations on: psychiatric facilities (Proshansky, Ittelson, & Rivlin, 1970); educational institutions (Barker & Gump, 1964; Stebbins, 1974); correctional institutions (Gill, 1962; Sommer, 1972; Turner, 1969); and whole communities (Barker, 1968; Barker & Wright, 1955).

The research reviewed represents only a beginning. Considerable work needs to be conducted to develop and refine a comprehensive technology for the behavioral assessment of living environments. In the future, behavioral scientists need to broaden their scope to evaluate how design changes in living environments are planned and implemented on a large scale. Land developers, urban architects, and city planners engage in behaviors that lead to major environmental modifications. These large-scale changes ultimately have the greatest behavioral impact of all.

Assessment of Community Problems

From the behavioral viewpoint, many community problems exist because community residents engage in behaviors that are inappropriate, but are tolerated by the community as a whole. Two examples of such behaviors are littering and refusing to conserve energy resources.

Littering

Several methods have been used to behaviorally assess littering. The most straightforward method is direct observation. For example, in one study (Kohlenberg & Phillips, 1973), trained observers kept records of the number of litter deposits made by patrons of a metropolitan zoo.

While intensive direct observation procedures produce precise and valuable data, they are impractical in most field applications. An alternative to direct observation is the assessment of the permanent products of littering. Ground survey techniques have been used for this purpose. In a ground survey, specified areas of land are marked off and counts are made of the number of pieces of litter present. Chapman and Risley (1974) used a ground survey to evaluate the effectiveness of antilitter procedures in a low-income urban housing project. Daily counts of litter were made under the following conditions: (1) a verbal appeal to neighborhood children to pick up litter; (2) payment of 10¢ per bag to children for filling bags with trash; (3) payment of 10¢ per yard cleaned. Payment for cleaning yards was the most effective procedure in reducing ground litter. Ground survey techniques have also been used to monitor littering in supervised forest campground areas (Clark, Burgess, & Hendee, 1972) and unsupervised forest campground areas (Powers, Osborne, & Anderson, 1973).

A second permanent product measure of littering is obtained by weighing trash collected in a given area. Measurements of litter weight take little time and can be made by untrained community residents. Burgess, Clark, and Hendee (1971) used a weight measure to study littering in neighborhood theaters during Saturday matinees. Usherettes collected and weighed litter deposited in trash cans, and on the floor of the theater. Providing children with incentives for depositing litter in trash cans resulted in appropriate deposit of 90% of the litter collected.

Although measurements of the weight of litter are practical means of assessment, they are not without problems. Hayes, Johnson, and Cone (1975) point out that reinforcing community residents for the amount of litter they deposit may have unintended results. Residents may import trash or they may collect only large pieces of litter because the payoff is greater. Hayes et al. (1975) maintain that the *marked item technique* overcomes these problems. In the marked item technique, specially marked pieces of litter are planted in target areas. These items are marked unobtrusively so that they can be identified by those in charge of antilitter programs. Community residents who collect litter are given bonuses for each marked item found in the trash they turn in. Ground surveys conducted to validate the marked item technique reveal that providing incentives for the collection of marked items reduces the total amount of ground litter (Hayes et al., 1973).

The marked item technique is an assessment procedure that is potentially applicable to large community systems, such as housing developments, parks, or even entire cities or towns (Hayes et al., 1975).

> The use of clear fluorescent paint and a black light, isotopes and a geiger counter, magnetic fluid or tape and a steel collector, or other materials enabling the mass screening of litter should make a very large-scale project feasible. If such a screening method could be found, it might be possible to automate the screening of trash and delivery of the reinforcer. Perhaps several automatic bins could be placed in a city, for example, and reinforcement delivered mechanically for bags of trash containing the specially marked items. (p. 385)

The costs of such a project are easily controlled by limiting the number of marked items distributed over an area. The use of such automated recording devices on a large scale remains to be demonstrated.

Conservation of Energy Resources

With the advent of the energy crisis, conservation is a critical issue. Behavioral assessment provides one means for evaluating the effectiveness of community-based conservation strategies.

One way to conserve resources is to reduce consumption. Kohlenberg, Phillips, and Proctor (1976) studied several techniques that may be used to modify patterns of electrical energy consumption. Recording devices installed in the homes of three volunteer families measured the amount of electricity used. Providing families with feedback regarding their electrical usage or incentives for reducing usage during peak energy demand periods led to improved patterns of energy conservation. The conclusion drawn from this study should be tentative; while feedback or incentives may lead to energy conservation in volunteer families, the effects of these procedures in a broader segment of society are not known.

A second way to conserve natural resources is through recycling. Many community members avoid recycling and buy only goods that are disposable. Geller, Farris, and Post (1973) studied several interventions aimed at prompting consumers to buy soft-drink bottles that could be recycled. The setting for the study was a small grocery store. A cashier and trained observer recorded whether customers purchased soft drinks in returnable or nonreturnable bottles. Prompting consumers with handbills urging them to buy drinks in returnable rather than nonreturnable bottles increased the number of returnable-bottle customers by an average of 25% compared to a baseline control periods.

Further research is needed to assess the long-term effects of such prompting procedures. The schedules under which prompts are delivered is likely to be an important variable. Consumers probably become satiated if prompts are given each time they visit a store, whereas intermittent prompting may lead to more permanent behavior change.

The underutilization of mass transportation facilities contributes to the present energy crisis as well. Gasoline consumption could be significantly reduced if consumers made greater use of existing mass transit. A recent study by Everett, Hayward, and Meyers (1974) examined the problem of how to increase bus ridership. Two buses equipped with turnstile counters were targeted for study. During baseline, 200 to 300 riders used each bus daily. During the experimental phase of the project, tokens exchangeable for a variety of reinforcers (free bus rides, food, play tickets, etc.) were given to all riders boarding the experimental bus, but not given to riders who used the control bus. Bus ridership on the experimental bus increased to 150% of baseline.

This study reveals the potential for relying on community residents to deal with community problems. Data on bus ridership was gathered by bus drivers. Tokens were dispensed by a single experimenter, but could easily have been dispensed by a mechanical device attached to the bus turnstile. Local businessmen allowed tokens to be exchanged for goods in their stores. The net result was a community system of mutually reinforcing interactions. Bus riders were reinforced for using the bus, the bus company was reinforced by increased ridership, and local merchants were reinforced by the increased business. Further research is needed to investigate the viability of this approach in larger mass transit systems.

Assessment of Social and Legal Reforms

Evaluating the impact of large-scale social interventions is a very difficult task. In large social systems, the behavioral scientist seldom has the luxury of precise control. It is difficult to attain the degree of rigor possible in true experiments. As a result, behavioral scientists have come to rely on quasi-experimental designs (Campbell & Stanley, 1966). These designs are only justified when better designs are not feasible. They permit the behavioral scientist to rule out some but not all plausible rival hypotheses that might explain the data. Because they are not true experiments, it is especially important that those who use them be thoroughly familiar with their weaknesses and limitations.

Ross, Campbell, and Glass (1970) examined the effects of a national law-enforcement crackdown on drunken drivers in Britain, using a quasi-experimental design. Well-publicized legislation was enacted which gave police the right to conduct on-the-scene Breathalyzer tests to assess blood alcohol levels of suspected drunken drivers. Data on fatalities and serious accident casualties were gathered before, during, and after onset of the crackdown and revealed that the crackdown had a maximal effect on fatalities and casualties occurring during the hours when British pubs and bars were open, a time when accidents were most likely to be affected by drinking. Campbell and Ross (1968) used a similar comparison control to assess the impact of a police crackdown on speeders in Connecticut; data on highway death rates in Connecticut were compared to data gathered from similar and adjacent states where no such crackdown was in effect. Obviously, such comparisons never approach the control possible in true experiments.

Quasi-experimental designs have also been employed to assess effects of legal innovations, such as compulsory arbitration (Rose, 1952; Stieber, 1949) or changes in capital punishment laws (Schuessler, 1969).

To summarize, a methodology is available that permits the behavioral assessment of social and legal interventions. It is best suited to the analysis of social interventions that have an abrupt, strong point of impact that can be dated (Ross et al., 1970). Gradual changes in policy are more difficult to assess using the methods discussed. Second, quasi-experimental design makes use of data gathered routinely by social agencies. This reduces the obtrusiveness and bias in comparison to procedures such as direct observation. Third, the methodology is applicable to retrospective analysis. This is a major advantage in field applications in which the behavioral scientist is often called in after the fact.

In dealing with large social systems, one needs to be keenly aware of political realities. It is often politically unrealistic to begin assessing social programs in a way that permits true experimental analysis. To funding agencies, the costs of such an enterprise seem too high and the benefits too low. In such cases, the benefits of behavioral assessment are best realized by *shaping* the system toward greater degrees of experimental rigor. Quasi-experimental designs are used initially; more sophisticated, true experimental designs are used later. Whatever type of design is used, it is incumbent upon the behavioral scientist to recognize its limitations and to frame his conclusions accordingly.

Issues of Broad Application

As behavioral assessment is more broadly applied, new and important issues must be faced. In this section, we consider theoretical, practical, and ethical issues.

Theoretical Issues

Behavioral analysis was originally developed to increase our understanding and ability to control the behavior of individual organisms. As this approach has been applied to larger community and societal systems, questions have been raised as to its theoretical adequacy.

Many of the very basic questions facing behavioral scientists who operate in community settings can not be answered by reference to the theoretical tenets of behavior therapy. Repucci and Saunders (1974) discuss this problem:

> ... Behavior modification was never intended to be a basis for describing, understanding, or changing natural settings. For example, there is nothing in behavior modification that guides us in answering these questions: Where should one seek to enter a setting? Where will the points of conflict arise? What will constitute a viable support system? What is a realistic time perspective for change? (p. 660)

Behavioral scientists recognize the need to explore alternative theoretical frameworks for answers to these questions (Franks & Wilson, 1975; Atthowe, 1973a; Kagel & Winkler, 1972; Winnett, 1974; Willems, 1974; Hetman, 1973). Such efforts are likely to yield new ways of conceptualizing and assessing problems of society and systems.

Practical Issues

Repucci and Saunders (1974) have listed eight problems facing those who apply behavioral principles in community settings. These problems are not unique to one community or social system. Because they are so pervasive, one needs to be thoroughly familiar with the limitations they impose on assessment efforts.

The Problem of Institutional Constraints

Schools, industries, local governments, and other established community systems operate according to well-established administrative procedures. Bureaucratic policies place limitations on what the behavioral clinician can do in the process of behavioral assessment. For example, resistance from residents may make it impossible to conduct direct observations in the community. "Red tape" may restrict access to important archival records. If the behavioral scientist is to be effective, he must realistically appraise the institutional constraints imposed upon him.

The Problem of External Pressure

No community system exists in a vacuum. Systems influence other systems. Social systems are particularly susceptible to pressures imposed by larger political and economic systems. The behavioral scientist does not have control over these sources of pressure. They may work for or against him. For example, fear of media exposure may make management reluctant to use certain assessment procedures to evaluate worker performance. External pressure can be very powerful. It can effectively stop ongoing programmatic assessment efforts. For this reason, attempts should be made to anticipate and circumvent problems due to external pressure.

The Problem of Two Populations

Another problem in dealing with community and societal systems is that two populations need to be considered simultaneously. One population is the target of the assessment efforts, e.g., mass transit riders, zoo patrons, or homeowners. A second population is made up of mediators indigenous to the system who gather the data and execute planned interventions. This second population includes, for example, usherettes who gather data on littering of theatergoers (Burgess, Clark, & Hendee, 1971), bus drivers who collect information on bus riders (Everett, Hayward, & Meyers, 1974), and teachers who monitor interracial interactions (Hauserman, Waley, & Behling, 1973). Assessment efforts need to focus not only on the behavior of the target population, but also on the behavior of these indigenous mediators. Without their assistance, assessment at a societal or systems level is unfeasible. The behavioral scientist must be ready to make judicious use of available reinforcers, such as letters of commendation, newspaper publicity, and opportunities to participate in professional presentations to gain support of mediators in the community.

The Problem of Language

The lack of common vocabulary between community members and the behavioral scientist may cause practical problems. Community residents may not understand terms such as discriminative stimulus, negative reinforcement, and multiple baseline. On the other hand, the language used by community residents may not be fully understood by the behavioral scientist. Language problems are notoriously pernicious. Neither party may become aware of them until they are glaringly apparent. The best safeguards against language problems are: (1) use plain, concrete terms; and (2) make frequent checks to ensure that terms are understood.

The Problem of Limited Resources

The resources available for behavioral assessment in applied settings are often limited. Typically, little or no money is available to support research assistants, to train observers, or to purchase automatic recording devices. Limited resources force the behavioral practitioner to rely on less expensive measurement methods and less rigorous experimental designs. While one must respond to the realities of applied settings, the limitations of such practical alternatives need to be fully recognized and appreciated.

The Problem of Labeling

Labels given to particular activities in community systems often influence the assessment process. Community planners are reluctant to consider assessing activities that are labeled "personality enriching," "recreational," "therapeutic," or "educational." Activities are frequently given a valued status simply because of their labels and not because of their actual function. The applied behavioral scientist needs to be aware of the nature and function of labels operating in systems he studies.

The Problem of Perceived Inflexibility
A common problem for behavioral scientists is that they are seen as more hard-nosed, objective, and rigid than other community members. While these perceptions may be incorrect (from the consultant's viewpoint), their influence is far-reaching. When dealing with community systems, the consultant needs to be aware of his public image. Systematic behavioral assessment programs are more likely to be accepted and supported if the consultant projects an open and reasonably flexible image.

The Problem of Compromise
In trying to gain acceptance and support for planned programs, the behavioral scientist changes his own behavior. He makes workable compromises. His behavior comes under the control and influence of values shared by the systems in which he works. Before getting involved in a system, the behavioral scientist must evaluate whether he agrees or disagrees with the values of that system.

Ethical Issues

Controversial ethical issues arise when behavioral assessment is applied to society and systems (Franks & Wilson, 1975; Bandura, 1974; Goldiamond, 1974).

One set of ethical questions is raised in the early stages of assessment. For example, when working in a large social system, whose interests are to be served? Does the behavioral scientist owe his allegiance to labor or management, to ghetto residents or City Hall? Who sets the behavioral goals for community and societal systems? Are they set by a high-ranking administrator, by elected officials, or by the segment of society that has the most political power? A second set of ethical questions is raised in the later stages of assessment, after information has been gathered and analyzed. How are data to be used to make more rational policy decisions? What is the behavioral scientist's role in social planning?

Up to the present time, behavioral scientists have avoided many of these troublesome ethical questions by restricting their efforts to target behavior that concern most members of the populace. Obviously, the interests of society are served by programs designed to alleviate widespread unemployment (Jones & Azrin,

1973); pollution (Kohlenberg & Phillips, 1973; Chapman & Risley, 1974; Clark, Burgess, & Hendee, 1972; Powers, Osborne, & Anderson, 1973); or crime (Schnelle, Kirchner, McNees, & Lawler, 1975). But what if the focus of behavioral assessment is shifted to behaviors whose relevance is questioned? In such cases, whose interests are to be served? These are not questions for the future. They are issues that are relevant whenever behavioral scientists deal with such complex issues as racial integration (Hauserman, Waley, & Behling, 1973) or birth control (Zifferblatt & Hendricks, 1974).

As the frontiers of behavioral assessment expand, we are faced with increasingly complex moral and ethical issues. The complexity of these issues should not serve as a deterrent to their being addressed. The danger that behavioral analysis can become a tool of a privileged, powerful few for the purpose of large-scale social intervention has been discussed (Franks & Wilson, 1975). Recognition and open discussion of the ethical obligations of applied behavioral analysis is an important first step in keeping this danger from becoming a reality. There are certain dangers involved in implementing large-scale changes in human behavior. Some of these are known and can be anticipated, others are unknown and cannot be anticipated. In determining the utility of behavioral assessment for social systems, one needs to weigh the costs of modifying behavior against the costs of not modifying behavior (Baer, 1974). A failure to assess and implement needed social changes may have major consequences for the quality of life and possibly even the survival of individuals, communities, and society itself.

Conclusions

Societal problems pose a real challenge to mankind. The behavioral approach to these problems is a promising one. The studies reviewed in Chapter 8 suggest that behavioral solutions to some social problems can be found. The theory, research, and practice of behavioral assessment at the level of social systems, however, are in their infancy. The research efforts reviewed represent only a beginning. The issues and problems inherent in this type of assessment are exceedingly complex. The assessment methods that have been used are exciting and innovative, but largely untested on a broad scale. A great deal of work needs to be done before behavioral assessment at the level of society and systems becomes a practical and scientifically sound reality.

CHAPTER 9

Future Directions

Behavioral assessment has developed beyond infancy and certainly is into its childhood. The unsettling years of adolescence lie ahead. The principles and procedures outlined in this book undoubtedly will be modified and refined in the years to come. What changes can be expected to occur? What directions will behavioral assessment take in the future? Chapter 9 attempts to answer these questions. We consider developments in three areas of behavioral assessment: teaching, service, and research.

Teaching Behavioral Assessment

Teaching Behavioral Assessment in Graduate Programs

During the 1960s and 1970s, formal graduate courses on traditional assessment have been on the decline. During the same time period, there has been a corresponding increase in the number of graduate and undergraduate courses in behavior therapy. Benassi and Lanson (1972) found a marked increase from 1955 to 1972 in the number of schools offering a first course in behavior therapy (see Figure 9.1).

Interest in training in behavior therapy has not been restricted to the field of psychology. The importance of training in behavior therapy in the fields of medicine (Agras, 1971), psychiatry (Edwards, Allen, & Verma, 1971; Gelfand, 1972), and social work (Shorkey, 1973) has been recognized. Separate courses in behavioral assessment have only begun to be developed (Evans & Nelson, 1974). One question likely to be explored and debated in the near future is: "What should be the content of a course on behavioral assessment?" At present, there are no definitive answers to

Figure 9.1 Cumulative Record by Year of Schools Offering Their First Course in Behavior Modification

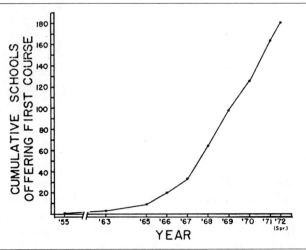

Reprinted from V. Benassi and R. Lanson. A survey of teaching of behavior modification in colleges and universities. *American Psychologist*, 1972, 27, 1063–1069. Copyright 1972 by the American Psychological Association. Reprinted by permission.

this question. Evans and Nelson (1974) outlined an approach to setting up a curriculum for a course in behavioral assessment. They stated:

> One challenge in constructing an assessment course is to reconcile the skills required for students' clinical work in the immediate future with a critical faculty needed to appreciate both the limitations of assessment and the corrective steps possible ... the concepts of reliability and validity were naturally omnipresent, but utility was heavily stressed.... The organizing principle was the single-subject research model, which integrated both the experimental study of psychopathology and the experimental analysis of individual behavior control and modification. (p. 599)

The Evans and Nelson article is highly recommended for those considering teaching behavioral assessment, both for its conceptualization and for its extensive bibliography. In our own teaching of behavioral assessment we have used the organizational structure shown in Table 9.1.

As more graduate programs begin to develop courses on behavioral assessment, there will be increasing diversity from program

TABLE 9.1
Outline for a Graduate-Level Course on Behavioral Assessment

Weeks & Topic

Introduction
1. Theoretical and Conceptual Base of Behavioral Assessment
2. Applied Behavior Analysis Research Designs
3. Behavioral Assessment Procedures and Methodological Problems

Applications
4. Anxiety
5. Depression
6. Case Presentations
7. Sexual and Marital Dysfunction
8. Interpersonal Social Behavior
9. Case Presentations
10. Children, Families, and School
11. Case Presentations
12. Institutional Settings
13. Society and Systems

Conclusions
14. Ethics and Values
15. Examination

to program. These differences will be a function of the program's focus, the beliefs and experiences of the instructors, and the purposes for which the students are being trained. Diversity is to be expected and, at this stage of development, even encouraged. A major precaution to be taken into account is that behavioral assessment should be taught within a context of empiricism. The field is in too early a stage of development to rigidify the techniques currently in use. Techniques are important, but even more so is a firm grasp of concepts and guiding principles.

Teaching Behavioral Assessment in Clinical Psychology Internships

Johnson and Bornstein (1974) surveyed 100 internships approved by the American Psychological Association and found that 65% of the programs had at least one behaviorally oriented psychologist on the staff, 49% offered didactic training in behavior therapy, and 87% provided supervision in behavior therapy. The unique characteristic of the internship as a setting which combines training with clinical service will facilitate the teaching of behavioral assessment. Trainees will have an opportunity to experiment with a variety of procedures under the supervision of an experienced behavior therapist.

Teaching Behavioral Assessment to Professionals in Areas Other than Mental Health

The teaching of behavioral assessment need not be restricted to an inpatient unit or a community mental health center setting. Businessmen, architects, city planners, administrators, lawyers, and politicians (the list could go on) would benefit from being taught the practical aspects of this approach. The use of operational definitions, multiple sources of data collection, and data-based decision-making would result in increased efficiency and greater accountability. This is an area that is very likely to be actively explored in the future.

Teaching Behavioral Assessment to the Consumer

Many benefits would accrue if behavioral assessment were made available to the consumer. By participating in courses at the junior college and adult-education levels, the public would be in a better position to solve problems and evaluate programs that are being paid for with their tax dollars.

The Format of Behavioral Assessment Courses

As behavioral assessment is taught in a wider variety of settings to a wider range of populations, more attention will be given to the format of such courses. The typical graduate-school format of a proseminar may be ineffective when applied in other settings. Individualized, as well as programmed, instruction in behavioral assessment is likely to be explored. New texts, teaching materials, workshops, etc., are likely to proliferate. Different members of the community (i.e., graduate students, lawyers, housewives) will require different types of materials.

Evaluation of Course Effectiveness

Courses on behavioral assessment need to be evaluated from the standpoint of what they purport to teach. Evaluation should make use of the assessment principles actually being taught: specification of the required skills, direct observation of students' behavior in analogue or natural settings, data collection, and feedback should be used extensively.

Service Directions

The combined factors of reduced use of traditional assessment procedures and increased applications of behavior therapy have apparently resulted in an assessment gap in applied settings. Behavioral assessment, which has the potential to fill this gap, will likely deal with several areas and issues in the future.

Mental Health Fields

Behavioral assessment has much to offer traditional mental health settings (e.g., hospitals, clinics, halfway houses) in terms of service. In the future, behavioral assessment procedures will find their way into the repertoire of more and more nonbehavioral clinicians. We have seen traditional family therapists ask parents to record the frequency of temper tantrums. Self-proclaimed eclectic therapists have been observed to have their clients monitor their child's temper tantrums, antecedents, and consequences. These few examples point to the growing trend for behavioral assessment to be used in mental health settings.

The creation of behavioral assessment units will become more commonplace. These units may be formed around specific problems, such as phobias, alcoholism, sexual dysfunctions, and depression, or around specific patient populations, such as children, adults, and geriatrics. They could be located in institutions or even in the community itself. Such units would have the resources to provide high-quality behavioral analyses. Multiple assessment strategies, too costly for most settings, could be used.

Behavioral assessment units have the added potential of using personnel with different levels of experience and training. Paraprofessionals who lack advanced formal training in the mental health field could be trained to perform many of the functions currently performed by doctoral-level personnel or not done at all for lack of manpower. Behavioral assessment units could use resources that are not otherwise available, such as a cadre of trained observers or psychophysiological instrumentation. A research orientation may well be expected in such a unit. It is probable that our most significant research findings regarding behavioral assessment will emerge from such service settings.

With the establishment of specialty units, referrals would come from clinicians who recognize the utility of the approach but lack the skill, motivation, or resources to use it themselves. The assessment unit would then share the results of the behavioral assessment with the referring clinician in some standardized format (e.g., providing specific behavioral data on observations, with norms and interpretative evaluations). This approach may be the most palatable in a clinical setting where there are many nonbehavioral clinicians. These clinicians may be more inclined to incorporate procedures of demonstrated utility.

Refinement of Procedures

With increased application, behavioral assessment procedures will be refined and made more efficient. Liberman, King, and DeRisi (1976) have demonstrated that practical procedures for assessment of patient and staff behavior can be developed for an entire community mental health center. Practical, cost-effective methods of assessment promise to replace such time-consuming procedures as full-scale psychological batteries.

Ethics

Ethical issues raised in behavioral assessment will likely be receiving increased attention. The emphasis on objectivity, clear statement of problems, and measurement will contribute to greater delineations of ethical responsibility. Thought will have to be given to the use of certain behavioral assessment procedures. Naturalistic observations are a case in point. The use of this procedure in closed settings raises questions as to the level of informed consent deemed appropriate, which parties will have access to what information, who "owns" the data, etc.

Legal Issues

The legal status of certain assessment techniques is already receiving increased attention. This trend is likely to continue. Is biofeedback apparatus illegal if used by a nonphysician? Do assertion techniques constitute an invasion of privacy? What levels of informed consent are required for which types of assessment procedures? Peer review boards and debates over licensure will become more prominent. Decisions derived from such discussions may determine which people with what qualifications are permitted to do what kinds of assessment. Future laws should serve to protect the individual client, the therapist, and society at large.

Technological Advances

Technological advances in the field of behavioral assessment are occurring at an astounding rate. In fact, by the time this book is published, many of the devices described may be outdated. This rapid pace of technological refinement and change is very likely to continue. Reductions in cost and size of equipment will enable us to collect data more efficiently. Implanted electrodes and telemetry devices may be used to monitor physiological responses such as heart rate, respiration, and body temperature of patients in a variety of natural settings. Advances in solid-state technology will lead to miniaturization of audio devices and transmitting equipment. Portable videotape units may be used to monitor individuals as they go through the day. Development of many of these techniques is already underway. As they are refined in the future they will increase the utility of behavioral assessment in providing direct patient service.

Research Directions

In the past, behavior therapy research focused on the development of various intervention techniques (e.g., systematic desensitization and covert sensitization). Now increasing emphasis is being placed on the evaluation of these procedures. As a result, behavioral assessment has become significant as an area of research in and of itself.

As behavioral assessment is applied more extensively, new questions undoubtedly will arise. Research addressing these questions may lead to the development of innovative models of human behavior. At the same time, basic research hopefully will provide the clinician with improved tools with which to perform clinical assessment. The interaction between clinical service and research is a major strength of the behavioral approach (Lazarus & Davison, 1971).

Recently there has been a great deal of research interest in the methodological problems inherent in behavioral assessment (e.g., Johnson & Bolstad, 1973; Lipinski & Nelson, 1974). This interest is likely to grow.

Conclusions

The above review outlines some of the future directions for teaching, service, and research in behavioral assessment. Exciting possibilities for the understanding, control, and prediction of behavior await professionals and consumers alike. These possibilities are sure to increase the effectiveness of our approaches.

However, a final cautionary note seems in order. It has been acknowledged previously that the successful implementation of behavior modification in natural settings (non-research-oriented human service settings) is dependent upon solutions to problems that are not directly related to theoretical issues of social learning (Repucci & Saunders, 1974). The front-line practitioner who relies solely on learning theory principles will find many roadblocks to well-designed intervention programs. The implementation of behavioral assessment must be viewed in the sociopsychological context in which it is to be used. Herein lies not only the challenge of the future, but the excitement as well.

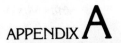

Family Questionnaire

Date:_____

Child's Name:_____ Address:_____
Age: _____ _____
Date of Birth:_____ Telephone:
School: _____ Home:_____
Grade: _____ Business :_____
Teacher: _____ School :_____

Questionnaire Completed By: _____
Relationship to Child: _____

I. **Home Information**

1. Father's Name:_____ Age:_____ Place of Birth:_____
   Occupation: _____ Work Hours:_____ Education_____
   Mother's Name:_____ Age:_____ Place of Birth:_____
   Occupation: _____ Work Hours:_____ Education_____

2. Language(s) Spoken at Home:_____

3. Brothers and Sisters:

Name:	Age:	Date of Birth:	School:	Grade:
_____	___	_____	_____	_____
_____	___	_____	_____	_____
_____	___	_____	_____	_____
_____	___	_____	_____	_____
_____	___	_____	_____	_____

4. Have these children ever had problems of a medical, academic, speech, hearing, visual or emotional nature? If so, please describe:
   _____
   _____
   _____
   _____

5. Other persons in the home:_____

6. Are you and your spouse the natural parents of this child and his/her siblings?_____
   _____

7. Is this your first marriage? _____ For your spouse?_____
8. Would you describe your relationship with your spouse as:

   Excellent_____ Good_____ Average_____ Fair_____ Poor_____

II. **Medical Information**
   1. Name and Address of Child's Physician:_____

   2. Current Medication (if any):_____

III. **Treatment (past and present)**
   1. Child has received special services at school (remedial reading, speech therapy, supplemental help, special class, psychological evaluation, etc.)

Type of Service	Dates Received	Name of Specialist
_____	_____	_____
_____	_____	_____
_____	_____	_____
_____	_____	_____

   2. Child has received special services privately (evaluation and or treatment: psychological, psychiatric, speech, hearing, visual, educational, medical)

Type of Service	Dates Received	Name of Specialist
_____	_____	_____
_____	_____	_____
_____	_____	_____
_____	_____	_____

IV. **Statement of the Problem**
   1. Parental statement of problem:_____

   2. When first noticed?_____

   3. What do you think was (is) the cause(s)?_____

Appendix A

4. Does the problem vary, being worse at certain times than others?
   _____
   _____
   _____
   _____

5. Do you and your spouse agree on the nature of your child's problem?
   _____
   _____
   _____
   _____

6. Do you and your spouse agree on child rearing, discipline, etc.? Explain.
   _____
   _____
   _____
   _____

7. Briefly describe how you and your spouse get along with this child.
   _____
   _____
   _____
   _____

8. Is the child having difficulty at school? If so, please describe:
   _____
   _____
   _____
   _____

9. What benefits do you hope to gain from therapy? _____
   _____
   _____
   _____
   _____

APPENDIX B
Multimodal Life History Questionnaire

PURPOSE OF THIS QUESTIONNAIRE:

THE PURPOSE OF THIS QUESTIONNAIRE IS TO OBTAIN A COMPREHENSIVE PICTURE OF YOUR BACKGROUND IN SCIENTIFIC WORK, RECORDS ARE NECESSARY, SINCE THEY PERMIT A MORE THOROUGH DEALING WITH ONE'S PROBLEMS. BY COMPLETING THESE QUESTIONS AS FULLY AND AS ACCURATELY AS YOU CAN, YOU WILL FACILITATE YOUR THERAPEUTIC PROGRAM. YOU ARE REQUESTED TO ANSWER THESE ROUTINE QUESTIONS IN YOUR OWN TIME INSTEAD OF USING UP YOUR ACTUAL CONSULTING TIME.

IT IS UNDERSTANDABLE THAT YOU MIGHT BE CONCERNED ABOUT WHAT HAPPENS TO THE INFORMATION ABOUT YOU BECAUSE MUCH OR ALL OF THIS INFORMATION IS HIGHLY PERSONAL. CASE RECORDS ARE STRICTLY CONFIDENTIAL. **NO OUTSIDER IS PERMITTED TO SEE YOUR CASE RECORD WITHOUT YOUR PERMISSION.**

IF YOU DO NOT DESIRE TO ANSWER ANY QUESTIONS, MERELY WRITE "DO NOT CARE TO ANSWER".

* *

Date _____

1. General
    Name: _____
    Address: _____

    Telephone numbers: (day) _____ (evenings) _____
    Age: _____ Occupation _____
    By whom were you referred? _____
    With whom are you now living (List people)? _____

    Do you live in a house, hotel, room, apartment, etc? _____
    Marital status: (Circle answer)
        Single; engaged; married; remarried; separated; divorced; widowed.
    If married, husband's (or wife's) name, age, occupation?
    Religion: a) In childhood
              b) As an adult

2. Clinical
    a) State in your own words the nature of your main problems and their duration:

    b) Give a brief account of the history and development of your complaints (from onset to present):

Copyright © 1977 Arnold A. Lazarus, Ph.D. All rights reserved. No part of this material may be reproduced by mimeograph, photocopying, or any other means without the written permission of the publisher. Published by: Multimodal Therapy Institute, P.O. Box 335, Kingston, New Jersey 08528.

Appendix B

c) On the scale below please estimate the severity of your problem(s):

Mildly Upsetting	Moderately Severe	Very Severe	Extremely Severe	Totally Incapacitating

d) Whom have you previously consulted about your present problem(s)?

e) Are you taking any medication? If "yes" what, how much, and with what results?

3. Personal Data:
   a) Date of birth:
   b) Mother's condition during pregnancy (as far as you know):
   c) Underline any of the following that applied during your childhood:

Night terrors	Bedwetting	Sleepwalking
Thumb sucking	Nail biting	Stammering
Fears	Happy childhood	Unhappy childhood

   Any others: _____

   c) Health during childhood?   Poor   Good   Excellent
   List illnesses: _____
   e) Health during adolescence?   Poor   Good   Excellent
   List illnesses: _____
   f) What is your height? _____
   g) Any surgical operations? (Please list them and give age at the time)

   h) Any accidents?

   i) List your five main fears:
   1.
   2.
   3.
   4.
   5.

   j) <u>Underline</u> any of the following that apply to you:

Headaches	Dizziness	Fainting spells
Palpitations	Stomach trouble	Anxiety
Bowel disturbances	Fatigue	No appetite
Anger	Take sedatives	Insomnia

Nightmares	Feel panicky	Alcoholism
Feel tense	Conflict	Tremors
Depressed	Suicidal ideas	Take drugs
Unable to relax	Sexual problems	Allergies
Don't like weekends & vacations	Over-ambitious	Shy with people
Can't make friends	Inferiority feelings	Can't make decisions
Can't keep a job	Memory problems	Home conditions bad
Financial problems	Lonely	Unable to have a good time
Excessive sweating	Often use aspirin or painkillers	Concentration difficulties
Others:		

k) <u>Underline</u> any of the following words that apply to you:

Worthless, useless, a "nobody", life is empty, inadequate, stupid, incompetent, naive, "can't do anything right", guilty, evil, morally wrong, horrible thoughts, hostile, full of hate, anxious, agitated, cowardly, unassertive, panicky, aggressive, ugly, deformed, unattractive, repulsive, depressed, lonely, unloved, misunderstood, bored, restless, confused, unconfident, in conflict, full of regrets, worthwhile, sympathetic, intelligent, attractive, confident, considerate
Others:

    l) Present interests, hobbies and activities: _____
_____

    m) How is most of your free time occupied? _____
    n) What is the last grade of schooling that you completed? _____
    o) Scholastic abilities; strengths and weaknesses: _____
_____

    p) Were you ever bullied or severly teased? _____
    q) Do you make friends easily? _____
    r) Do you keep them? _____

4. <u>Occupational Data</u>:
    a) What sort of work are you doing now?
    b) Kinds of jobs held in the past?
    c) Does your present work satisfy you? (If not, in what ways are you dissatisfied?)

    d) What do you earn? _____ How much does it cost you to live? _____
    e) <u>Ambitions</u>
        Past: _____
        Present: _____

Appendix B

5. <u>Sex Information</u>
   a) Parental attitudes to sex (e.g., was there sex instruction or discussion in the home?)

   b) When and how did you derive your first knowledge of sex?

   c) When did you first become aware of your own sexual impulses?

   d) Did you ever experience any anxieities or guilt feelings arising out of sex or masturbation? If "yes" please explain:

   e) Any relevant details regarding your first or subsequent sexual experience:

   f) Is your present sex life satisfactory? (if not, please explain)

   g) Provide information about any significant heterosexual (and/or homosexual) reactions:

6. <u>Menstrual History</u>:
   Age at first period? _____ Were you informed or did it come as a shock? _____
   Are you regular? _____ Duration _____ Do you have pain? _____
   Date of last period: _____ Do your periods affect your moods? _____

7. <u>Marital history</u>:
   How long did you know your marriage partner before engagement? _____
   Husband's / Wife's Age: _____ Occupation of husband or wife: _____
   a) Personality of husband or wife (in your own words):

   b) In what areas is there compatibility?

   c) In what areas is there incompatibility?

   d) How do you get along with your in-laws? (this includes brothers and sisters-in-law)

   How many children have you? _____
   Please list their sex and age(s)

e) Do any of your children present special problems?

f) Any relevant details regarding miscarriages or abortions?

g) Comments about any previous marriage(s) and brief details.

8. Family Data:
   a) Father:
      Living or deceased?
      If deceased, your age at the time of his death?
      Cause of death?
      If alive, father's present age?
      Occupation:
      Health:
   b) Mother:
      Living or deceased?
      If deceased, your age at the time of her death?
      Cause of death?
      If alive, mother's present age?
      Occupation:
      Health:
   c) Siblings
      Number of brothers: Brothers Ages:
      Number of sisters: Sisters Ages:
   d) Relationship with brothers and sisters:
      a) Past

      b) Present

   e) Give description of your father's personality and his attitude towards you (past and present):

Appendix B

f) Give a description of your mother's personality and her attitude towards you (past and present):

g) In what ways were you punished by your parents as a child?

h) Give an impression of your home atmosphere (i.e., the home in which you grew up. Mention state of compatibility between parents and between children).

i) Were you able to confide in your parents?

j) Did your parents understand you?

k) Basically, did you feel loved and respected by your parents?

If you have a step-parent, give your age when parent remarried:

l) Give an outline of your religious training:

m) If you were not brought up by your parents, who raised you, and between what years?

n) Has anyone (parents, relatives, friends) ever interfered in your marriage, occupation, etc?

o) Who are the most important people in your life?

p) Does any member of your family suffer from alcoholism, epilepsy, or anything can be considered a "mental disorder?"

q) Are there any other members of your family about whom information regarding illness, etc., is relevant?

8A Additional information

a) Recount any fearful or distressing experiences not previously mentioned:

b) List the benefits you hope to derive from therapy.

c) List any situations that make you feel calm or relaxed.

d) Have you ever lost control (e.g., temper or crying or aggression?) If so, please describe.

e) Please add any information not tapped by this questionnaire that may aid your therapist in understanding and helping you.

f) Have you ever attempted or seriously comtemplated suicide? If so please describe

9. Self-Description: (Please complete the following:)

a) I am a person who _____
b) All my life _____
c) Ever since I was a child _____
d) One of the things I feel proud of is _____
e) It's hard for me to admit _____
f) One of the things I can't forgive is _____
g) One of the things I feel guilty about is _____
h) A good thing about having problems is _____
i) One of the ways people hurt me is _____
j) I could shock you by _____
k) A mother should _____
l) If I get angry with you _____

Appendix B

   m) A father should _____
   n) If I weren't afraid to be myself, I might _____
   o) One of the things I'm angry about is _____
   p) If I told you what Im feeling now _____
   q) The bad thing about growing up is _____
   r) One of the ways I could help myself but don't is _____

10. **Assessment Summary**

   a) What is there about your present BEHAVIOR that you would like to change?

   b) What feelings do you wish to alter (e.g. increase or decrease)?

   c) What SENSATIONS are especially:
      1) Pleasant for you?

      2) Unpleasant for you?

   d) Describe a very pleasant IMAGE or fantasy

   e) Describe a very unpleasant IMAGE or fantasy

   f) What do you consider your most irrational THOUGHT or idea?

   g) Describe any interpersonal RELATIONSHIPS that give you:
      1) Joy

      2) Grief

   h) What personal characteristics do you think the ideal therapist should possess?

   i) How would you describe an ideal therapist's interactions with his / her clients?

   j) What do you think therapy will do for you and how long do you think your therapy should last?

   k) In a few words, what do you think therapy is all about?

11. Sequential History

Please outline your most significant memories and experiences within the following ages:

0 - 5 _____

6 - 10 _____

11 - 15 _____

16 - 20 _____

21 - 25 _____

26 - 30 _____

31 - 35 _____

36 - 40 _____

41 - 45 _____

46 - 50 _____

51 - 55 _____

56 - 60 _____

61 - 65 _____

Over 66 _____

12. Word Pictures

Use the remaining space to give a word - picture of yourself as would be described:

a) By yourself
b) By your spouse (if married)
c) By your best friend
d) By someone who dislikes you

References

Agras, W. S. Transfer during systematic desensitization therapy. *Behavior Research & Therapy,* 1967, 5, 193–200.
Agras, W. S. The role of behavior therapy in teaching medical students. *Journal of Behavior Therapy and Experimental Psychiatry,* 1971, 2, 219–222.
Agras, W. S. Behavior modification in the general hospital psychiatric unit. In H. Leitenberg (Ed.), *Handbook of behavior modification and behavior therapy.* Englewood Cliffs, N.J.: Prentice-Hall, 1976.
Aldis, O. Of pigeons and men. In J. Ulrich, T. Stachnik, & J. Mabry (Eds.), *Control of human behavior.* Vol. 1. Glenville, Ill.: Scott Foresman, 1966.
Allport, G. W. Traits revisited. *American Psychologist,* 1966, 21, 1–10.
Alper, T. G., & White, O. R. The behavior description referral form: A tool for the school psychologist in the elementary school. *Journal of School Psychology,* 1971, 9, 177–181.
Alpert, R., & Haber, R. Anxiety in academic achievement situations. *Journal of Abnormal and Social Psychology,* 1960, 61, 207–215.
Annon, J. S. *The behavioral treatment of sexual problems.* Vol. 1. *Brief therapy.* Honolulu: Enabling Systems, 1974.
Annon, J. S. *The behavioral treatment of sexual problems.* Vol. 2. *Intensive therapy.* Honolulu: Enabling Systems, 1975. (a)
Annon, J. S. *The sexual fear inventory—Female form.* Honolulu: Enabling Systems, 1975. (b)
Annon, J. S. *The sexual fear inventory—Male form.* Honolulu: Enabling Systems, 1975. (c)
Annon, J. S. *The sexual pleasure inventory—Female form.* Honolulu: Enabling Systems, 1975. (d)

Annon, J. S. *The sexual pleasure inventory—Male form.* Honolulu: Enabling Systems, 1975. (e)
Arkowitz, H., Lichtenstein, E., McGovern, & Hines, P. The behavioral assessment of social competence in males. *Behavior Therapy,* 1975, *6,* 3–13.
Arnkoff, D. B., & Stewart, J. The effectiveness of modeling and videotape feedback on personal problem solving. *Behavior Research & Therapy,* 1975, *13,* 127–134.
Atthowe, J. M. Behavior innovation and persistence. *American Psychologist,* 1973, *23,* 34–41. (a)
Atthowe, J. Token economies come of age. *Behavior Therapy,* 1973, *4,* 646–654. (b)
Ayllon, T., & Azrin, N. H. *The token economy: A motivational system for therapy and rehabilitation.* Englewood Cliffs, N.J.: Prentice-Hall, 1968.
Azrin, N. H., Holz, W., & Goldiamond, I. Response bias in questionnaire reports. *Journal of Consulting Psychology,* 1961, *25,* 324–326.
Azrin, N. H., & Nunn, R. G. A rapid method of eliminating stuttering by a regulated breathing approach. *Behavior Research & Therapy,* 1974, *12,* 279–286.
Azrin, N. H., & Powell, J. Behavioral engineering: The reduction of smoking behavior by a conditioning apparatus and procedure. *Journal of Applied Behavior Analysis,* 1968, *1,* 193–200.
Baer, D. M. A note on the absence of a Santa Claus in any known ecosystem: A rejoinder to Willems. *Journal of Applied Behavior Analysis,* 1974, *7,* 167–170.
Baker, B. L., Cohen, D. C., & Saunders, J. T. Self-directed desensitization for acrophobia. *Behavior Research & Therapy,* 1973, *11,* 79–90.
Bandura, A. *Principles of behavior modification.* New York: Holt, Rinehart and Winston, 1969.
Bandura, A. The ethics and social purposes of behavior modification. Paper presented at the eighth annual meeting of the Association for the Advancement of Behavior Therapy, Chicago, 1974.
Barker, R. *Ecological psychology.* Stanford, Cal.: Stanford University Press, 1968.
Barker, R. G., & Gump, P. V. *Big school, small school: High school size and student behavior.* Stanford, Cal.: Stanford University Press, 1964.

Barker, R. G., & Wright, H. F. *Midwest and its children.* Evanston, Ill.: Row, Peterson, 1955.
Barlow, D. H., & Hersen, M. Single-case experimental designs: Uses in applied clinical research. *Archives of General Psychiatry,* 1973, 29, 319–325.
Beck, A. T., *Depression: Clinical, experimental and theoretical aspects.* New York: Harper & Row, 1967.
Begelman, D. A. Ethical and legal issues of behavior modification. In M. Hersen, R. M. Eisler, & P. M. Miller (Eds.), *Progress in behavior modification:* Vol. 1. New York: Academic Press, 1975.
Benassi, V., & Lanson, R. A. Survey of the teaching of behavior modification in colleges and universities. *American Psychologist,* 1972, 27, 1063–1069.
Bernstein, D. A., & Beatty, W. E. The use of in vivo desensitization as part of a total therapeutic intervention. *Journal of Behavior Therapy and Experimental Psychiatry,* 1971, 2, 259–266.
Bijou, S. W., Peterson, R. F., Harris, F. R., Allen, K. E., & Johnston, M. S. Methodology for experimental studies of young children in natural settings. *Psychological Record,* 1969, 19, 177–210.
Birchler, G. R. Differential patterns of instrumental affiliative behavior as a function of degree of marital distress and level of intimacy. Unpublished doctoral dissertation, University of Oregon, 1972.
Blanchard, E. B., & Young, L. B. Clinical application of biofeedback training. *Archives of General Psychiatry,* 1974, 30, 573–589.
Bloch, J. P. Agreement between parents' and teachers' ratings of childhood emotional adjustment. Unpublished master's thesis, University of Louisville, 1971.
Borkovec, T. D. Heart-rate process during systematic desensitization and implosion therapy for analog anxiety. *Behavior Therapy,* 1974, 5, 636–641.
Borkovec, T. D., Kaloupek, D. G., & Sloma, K. M. The facilitative effect of muscle-tension release in the relaxation treatment of sleep disturbance. *Behavior Therapy,* 1975, 6, 301–309.
Brown, R. C. Instruments in psychophysiology. In N. S. Greenfield & R. A. Sternbach (Eds.), *Handbook of psychophysiology.* New York: Holt, Rinehart and Winston, 1972.
Brown, R. E., Copeland, R. E., & Hall, R. V. The school principal

as a behavior modifier. *Journal of Educational Research,* 1972, *66,* 175–180.
Browning, R. M., & Stover, D. O. *Behavior modification in child treatment.* Chicago: Aldine, 1971.
Budzynski, T. H., Stoyva, J. M., & Adler, C. Feedback-induced muscle relaxation: application to tension headache. *Journal of Behavior Therapy and Experimental Psychiatry,* 1970, *1,* 205–211.
Burgess, R. L., Clark, R. N., & Hendee, J. C. An experimental analysis of anti-litter procedures. *Journal of Applied Behavior Analysis,* 1971, *4,* 71–75.
Campbell, D. T., & Ross, H. The Connecticut crackdown on speeding; Time series data in quasi-experimental analysis. *Law and Society Review,* 1968, *8,* 33–53.
Campbell, D. T., & Stanley, J. C. *Experimental and quasi-experimental design for research.* Chicago: Rand McNally, 1966.
Cataldo, M. F., & Risley, T. R. Evaluation of living environments: The manifest description of ward activities. In P. O. Davidson, F. W. Clark, & L. A. Hamerlynck (Eds.), *Evaluation of behavioral programs in community, residential and school settings. Proceedings of the Fifth Banff International Conference on Behavior Modification.* Champaign, Ill.: Research Press, 1974.
Cattell, R. B. *Personality: A systematic theoretical and factual study.* New York: McGraw-Hill, 1950.
Cautela, J. R. Covert sensitization. *Psychological Record,* 1967, *20,* 459–468.
Cautela, J. R. Covert reinforcement. *Behavior Therapy,* 1970, *1,* 33–50.
Cautela, J. R., & Kastenbaum, R. A reinforcement survey schedule for use in therapy, training and research. *Psychological Reports,* 1967, *20,* 1115–1130.
Cautela, J., & Wisocki, P. The use of male and female therapists in the treatment of homosexual behavior. In R. Rubin & C. Franks (Eds.), *Advances in Behavior Therapy: 1968.* New York: Academic Press, 1969.
Cautela, J. R., & Upper, D. The process of individual behavior therapy. In M. Hersen, R. M. Eiseler, & P. M. Miller (Eds.), *Progress in behavior modification.* Vol. 1. New York: Academic Press, 1975.
Chapman, C., & Risley, T. R. Anti-litter procedures in an urban high-density area. *Journal of Applied Behavior Analysis,* 1974, *7,* 377–384.

Chesney, M. A., & Tasto, D. L. The development of the menstrual symptom questionnaire. *Behavior Research & Therapy,* 1975, *13,* 237–244.

Clark, R. N., Burgess, R. L., & Hendee, J. The development of anti-litter behavior in a forest campground. *Journal of Applied Behavior Analysis,* 1972, *5,* 1–6.

Conway, J. B., & Bucher, B. D. Transfer and maintenance of behavior change in children: A review and suggestions. In E. J. Mash, L. A. Hamerlynck, & L. C. Handy (Eds.), *Behavior modification and families.* New York: Brunner/Mazel, 1976.

Cooper, J. The Leyton Obsessional Inventory. *Psychological Medicine,* 1970, *1,* 48–64.

Cooper, K. H. *The new aerobics.* New York: Bantam Books, 1970.

Copeland, R. E., Brown, R. E., Axelrod, S., & Hall, R. V. Effect of a school principal praising parents for student attendance. *Educational Technology,* 1972, *12,* 56–59.

Copeland, R. E., Brown, R. E., & Hall, R. V. The effects of principal-implemented techniques on the behavior of pupils. *Journal of Applied Behavior Analysis,* 1974, *7,* 77–86.

Cronbach, L. J. *Essentials of psychological testing.* (2nd ed.) New York: Harper & Row, 1960.

Cummins, W. W. A bird's eye glimpse of men and machines. In J. Ulrich, T. Stachnik, & J. Mabry (Eds.), *Control of human behavior.* Vol. 1. Glenville, Ill.: Scott Foresman, 1966.

Curran, J. P., & Gilbert, F. S. A test of the relative effectiveness of a systematic desensitization program and interpersonal skills training program with data anxious subjects. *Behavior Therapy,* 1975, *6,* 510–521.

Daily, C. A. The practical utility of the clinical report. *Journal of Consulting Psychology,* 1953, *17,* 297–302.

Davison, G. C. Appraisal of behavior modification techniques with adults in institutional settings. In C. M. Franks (Ed.), *Behavior therapy: Appraisal and status.* New York: McGraw-Hill, 1969.

Department of Health, Education & Welfare. National Institute of Health. The institutional guide to DHEW policies on protection of human subjects. DHEW pub. no. (NIH) 72-102, 1971.

Diagnostic and statistical manual of mental disorders (2nd ed.—DSM-II). American Psychiatric Association, 1968.

Edwards, N. B., Allen, R., & Verma, S. Resident involvement in residency training. *Journal of Behavior Therapy and Experimental Psychiatry,* 1971, *2,* 303–306.

Ellis, A. *Reason and emotion in psychotherapy.* New York: Lyle Stuart, 1962.
Eisler, R. M., Miller, P., & Hersen, M. Components of assertive behavior. *Journal of Clinical Psychology,* 1973, *29,* 295–299.
Emmelkamp, P. M. G., & Ultee, K. A. A comparison of "successive approximation" and "self-observation" in the treatment of agoraphobia. *Behavior Therapy,* 1974, *5,* 606–613.
Endler, N. S., Hunt, J. McV., & Rosenstein, A. J. An S-R inventory of anxiousness. *Psychological Monographs,* 1962, *76* (Whole No. 536).
Epstein, L. E., & Peterson, G. L. Differential conditioning using covert stimuli. *Behavior Therapy,* 1973, *44,* 96–99.
Epstein, L. H., & Hersen, M. A multiple baseline analysis of coverant control. *Journal of Behavior Therapy and Experimental Psychiatry,* 1974, *5,* 7–12.
Ernst, F. A. Self-recording and counter-conditioning of a self-mutilative compulsion. *Behavior Therapy,* 1973, *4,* 144–146.
Evans, D. R., & Bond, I. K. Reciprocal inhibition therapy and classical conditioning in the treatment of insomniacs. *Behavior Research & Therapy,* 1969, *7,* 323–326.
Evans, I. M., & Nelson, R. O. A curriculum for the teaching of behavior assessment. *American Psychologist,* 1974, *29,* 598–606.
Evans, P. D., & Kellam, A. Semi-automated desensitization: A controlled clinical trial. *Behavior Research & Therapy,* 1973, *11,* 641–646.
Everaerd, W., Rijkin, H. M., & Emmelkamp, P. M. A comparison of "flooding" and "successive approximation" in the treatment of agoraphobia. *Behavior Research & Therapy,* 1973, *11,* 105–118.
Everett, P. B., Hayward, S. C., & Meyers, A. W. The effects of a token reinforcement procedure on bus ridership. *Journal of Applied Behavior Analysis,* 1974, *7,* 1–10.
Eyberg, S. M., & Johnson, S. M. Multiple assessment of behavior modification with families: Effects of contingency contracting and order of treated problems. *Journal of Consulting and Clinical Psychology,* 1974, *42,* 594–606.
Eysenck, H. J. *Handbook of abnormal psychology.* New York: Basic Books, 1961.
Eysenck, H. J. *Handbook of abnormal psychology.* London: Pitman Medical Publishers, 1973.
Fairweather, G. W. *Methods in experimental social innovations.* New York: John Wiley, 1967.
Fairweather, G. W., Sanders, D. H., Maynard, H., & Cressler, D.

H. *Community life for the mentally ill: An alternative to institutional care.* Chicago: Aldine, 1969.
Fay, A. F. The drug modality. In A. A. Lazarus (Ed.), *Multimodal behavior therapy.* New York: Springer Pub. Co., 1976.
Ferster, C. B. Classification of behavioral pathology. In L. Krasner & L. P. Ullmann (Eds.), *Research in behavior modification.* New York: Holt, Rinehart and Winston, 1965.
Forness, S. R., & Esveldt, K. Classroom observation of children referred to a child psychiatric clinic. *Journal of Child Psychiatry,* 1974, *13,* 335–343.
Franks, C. M., & Wilson, G. T. *Annual review of behavior therapy: Theory and practice, Volume 3: 1975.* New York: Brunner/Mazel, 1975.
Frederiksen, L. W. Treatment of ruminative thinking by self-inventory. *Journal of Behavior Therapy and Experimental Psychiatry,* 1975, *6,* 258–259.
Fryrear, J. L., & Weiner, S. Treatment of a phobia by use of a videotaped modeling procedure: A case study. *Behavior Therapy,* 1970, *1,* 391–394.
Furman, S. Intestinal biofeedback in functional diarrhea: A preliminary report. *Journal of Behavior Therapy and Experimental Psychiatry,* 1973, *4,* 317–322.
Gagnon, J. H. Scripts and the coordination of sexual conduct. Nebraska Symposium on Motivation, 1974.
Gagnon, J. H., & Simon, W. *The social sources of sexual conduct.* Chicago: Aldine, 1973.
Galassi, J. P., DeLo, J. S., Galassi, M. D., & Bastien, S. The college self-expression scale: A measure of assertiveness. *Behavior Therapy,* 1974, *5,* 165–171.
Gaupp, L. A., Stern, R. M., & Ratliff, R. G. The use of aversion-relief procedures in the treatment of a case of voyeurism. *Behavior Therapy,* 1971, *2,* 585–588.
Gelfand, D. M., & Hartmann, D. P. *Child behavior analysis and therapy.* New York: Pergamon Press, 1975.
Gelfand, S. A behavior modification training program for psychiatric residents. *Journal of Behavior Therapy and Experimental Psychiatry,* 1972, *3,* 147–152.
Geller, E. S., Farris, J. S., & Post, D. S. Prompting a consumer behavior for pollution control. *Journal of Applied Behavior Analysis,* 1973, *6,* 367–376.
Gill, H. B. Correctional philosophy and architecture. *Journal of Criminal Law, Criminology and Police Science,* 1962, *53,* 312–322.

Glasgow, R. E., & Arkowitz, H. The behavioral assessment of male and female social competence in dyadic heterosexual interactions. *Behavior Therapy*, 1975, 6, 488–498.

Glass, G. V., Tiao, G. C., & Maguire, T. O. Analysis of data on the 1900 revision of German divorce laws as a time-series quasi-experiment. *Law and Society Review*, 1971, 4, 539–562.

Goldfried, M. R., & Davison, G. C. *Clinical behavior therapy.* New York: Holt, Rinehart and Winston, 1976.

Goldfried, M. R., and D'Zurilla, T. J. A behavioral-analytic model for assessing competence. In C. D. Spielberger (Ed.), *Current topics in clinical and community psychology.* Vol. 1. New York: Academic Press, 1969.

Goldfried, M. R., & Kent, R. N. Traditional versus behavioral personality assessment: A comparison of methodological and theoretical assumptions. *Psychological Bulletin*, 1972, 77, 409–420.

Goldiamond, I. Toward a constructional approach to social problems: Ethical and constitutional issues raised by applied behavior analysis. *Behaviorism*, 1974, 2.

Gordon, S. B. Comment: Multiple assessment of behavior modification with families. *Journal of Consulting and Clinical Psychology*, 1975, 43, 917. (a)

Gordon, S. B. The responsive parenting class: A behavioral approach to working with parents in a community mental health center. Paper presented at the Annual Meeting of the New Jersey Psychological Association, Somerset, N.J., May 1975. (b)

Gordon, S. B. Training teachers in behavior modification. Responsive teaching vs. a control group. Paper presented at the Annual Meeting of the American Psychological Association, Washington, D.C., September 1976.

Gordon, S. B., & Keefe, F. J. Naturalistic observations of classroom behavior. Manuscript submitted for publication, 1977.

Gordon, S. B., Lerner, L. L., & Keefe, F. J. Responsive parenting: An approach to training parents of problem children. *American Journal of Community Psychology*, in press.

Gottman, J. M., & Leiblum, S. R. *How to do psychotherapy and how to evaluate it: A manual for beginners*, 1973.

Grant, R. L., & Maletzky, B. M. Application of the WEED system to psychiatric records. *Psychiatry in Medicine*, 1972, 3, 119–129.

Gray, B. Theoretical approximations of stuttering adaptation:

Statement of predictive accuracy. *Behavior Research & Therapy,* 1965, *3,* 221–228.
Graziano, A. M. *Behavior therapy with children.* Chicago: Aldine, 1975.
Hackman, A., & McLean, C. A comparison of flooding and thought stopping in the treatment of an obsessional neurosis. *Behavior Research & Therapy,* 1975, *13,* 263–270.
Hagen, R. L., Craighead, W. E., & Paul, G. L. Staff reactivity to evaluative observation. *Behavior Therapy,* 1975, *6,* 201–205.
Hall, J., & Baker, R. Token economy systems: Breakdown and control. *Behavior Research & Therapy,* 1973, *11,* 253–263.
Hall, R. V. *Managing behavior: Part 1.* Kansas City: H & H Enterprises, 1971.
Hannum, J. W., Thoreson, C. E., & Hubbard, D. P., Jr. A behavioral study of self-esteem with elementary teachers. In M. J. Mahoney & C. E. Thoreson (Eds.), *Self-control: Power to the person.* Monterey, Cal.: Brooks-Cole, 1974.
Harmatz, M. G., Mendelsohn, R., & Glassman, M. L. Gathering naturalistic, objective data on the behavior of schizophrenic patients. *Hospital and Community Psychiatry,* 1975, *26,* 83–86.
Hartman, W. E., & Fithian, M. A. *Treatment of sexual dysfunction.* Long Beach, Cal.: Center for Marital and Sexual Studies, 1972.
Hauserman, N., Waley, S. R., & Behling, M. Reinforced racial integration in the first grade: A study in generalization. *Journal of Applied Behavior Analysis,* 1973, *6,* 193–200.
Hayes, S. C., Johnson, V. S., & Cone, J. D. The marked item technique: A practical procedure for litter control. *Journal of Applied Behavior Analysis,* 1975, *8,* 381–386.
Hayes-Roth, F., Longabaugh, R., & Ryback, R. The problem-oriented medical record and psychiatry. *British Journal of Psychiatry,* 1972, *121,* 27–34.
Heller, R. F., & Strang, H. R. Controlling bruxism through automated aversive conditioning. *Behavior Research & Therapy,* 1973, *11,* 327–330.
Herman, J. A., deMontes, A. J., Dominguez, B., Montes, F., & Hopkins, B. L. Effects of bonuses for punctuality on the tardiness of industrial workers. *Journal of Applied Behavior Analysis,* 1973, *4,* 563–572.
Hersen, M. Self-assessment of fear. *Behavior Therapy,* 1973, *4,* 241–257.

Hetman, F. Society and the assessment of technology. Paper presented at the Organization for Economic Cooperation and Development, Paris, France, May 1973.
Hodgson, R., & Rachman, S. The effects of contamination and washing in obsessional patients. *Behavior Research & Therapy,* 1972, *10,* 111–117.
Jacobson, N. S. Problem solving and contingency contracting in the treatment of marital discord. *Journal of Consulting and Clinical Psychology,* 1977, *45,* 92–100.
Jacobson, N. S., & Martin, B. Behavioral marriage therapy: Current status. *Psychological Bulletin,* 1976, *83,* 540–556.
Jason, L. Rapid improvement in insomnia following self-monitoring. *Journal of Behavior Therapy and Experimental Psychiatry,* 1975, *6,* 349–350.
Johansson, S., Lewinsohn, P. M., & Flippo, J. R. An application of the Premack Principle to the verbal behavior of depressed subject. Paper presented at the meeting of the Association for the Advancement of Behavior Therapy, April 1969.
Johnson, J. H., & Bornstein, P. H. A survey of behavior modification training opportunities in APA-approved internship facilities. *American Psychologist,* 1974, *29,* 342–348.
Johnson, R. K., & Meyer, R. G. Phased biofeedback approach for epileptic seizure control. *Journal of Behavior Therapy and Experimental Psychiatry,* 1974, *5,* 185–188.
Johnson, S. M., & Bolstad, O. D. Methodological issues in naturalistic observation: Some problems for field research. In L. A. Hamerlynch, L. C. Handy, & E. J. Mash (Eds.), *Behavior change: Methodology, concepts and practice.* Champaign, Ill.: Research Press, 1973.
Johnson, S. M., & Christensen, A. Multiple criteria follow-up of behavior modification with families. *Journal of Abnormal Child Psychology,* in press.
Johnson, S. M., Christensen, A., and Bellamy, G. T. Evaluation of family intervention through unobtrusive audio recordings: experiences in "bugging" children. *Journal of Applied Behavior Analysis,* 1976, *9,* 213–219.
Johnson, S. M., & Eyberg, S. M. Evaluating outcome data: A reply to Gordon. *Journal of Consulting and Clinical Psychology,* 1975, *43,* 917–919.
Johnson, S. M., Wahl, G., Martin, S., & Johanson, S. How deviant is the normal child? A behavioral analysis of the preschool child and his family. In R. D. Rubin, J. P. Brady, & J. D.

Henderson (Eds.), *Advances in behavior therapy.* New York: Academic Press, 1973.

Johnson, W. G. Behavior therapy—what place in the psychiatric residency? *Journal of Behavior Therapy and Experimental Psychiatry,* 1972, *3,* 329–332.

Jones, R. J., & Azrin, N. H. An experimental application of a social reinforcement approach to the problem of job-finding. *Journal of Applied Behavior Analysis,* 1973, *6,* 345–354.

Kagel, J. H., & Winkler, R. C. Behavioral economics: Areas of cooperative research between economics and applied behavioral analysis. *Journal of Applied Behavior Analysis,* 1972, *5,* 335–342.

Kanfer, F. H. Report on outcome measures in behavior therapy. In I. E. Waskow & M. B. Parloff (Eds.), *Psychotherapy change measures.* (DHEW Publication No. (ADM) 74-120) Washington, D.C.: U.S. Government Printing Office, 1975.

Kanfer, F. H., & Phillips, J. S. *Learning foundations of behavior therapy.* New York: John Wiley, 1970.

Kanfer, F. H., & Saslow, G. Behavioral analysis. *Archives of General Psychiatry,* 1965, *12,* 529–538.

Kanfer, F. H., & Saslow, G. Behavioral diagnosis. In C. M. Franks (Ed.), *Behavior therapy: Appraisal and status.* New York: McGraw-Hill, 1969.

Kaplan, H. S. *The new sex therapy.* New York: Brunner/Mazel, 1974.

Katchadourian, H. A., & Lunde, D. T. *Fundamentals of human sexuality.* New York: Holt, Rinehart and Winston, 1972.

Katz, D. An automated system for eliciting and recording self-observation during dyadic communication. *Behavior Therapy,* 1974, *5,* 689–697.

Kau, M. L., & Fischer, J. Self-modification of exercise behavior. *Journal of Behavior Therapy and Experimental Psychiatry,* 1974, *5,* 213–214.

Kazdin, A. E. Self-monitoring and behavior change. In M. J. Mahoney & C. E. Thoreson (Eds.), *Self-control: Power to the person.* Monterey, Cal.: Brooks-Cole, 1974.

Kazdin, A. E., & Kopel, S. A. On resolving ambiguities of the multiple-baseline design: Problems and recommendations. *Behavior Therapy,* 1975, *6,* 601–608.

Keefe, F. J., & Sirota, A. How objective is behavioral assessment? An historical analysis. Unpublished manuscript, Harvard Medical School, 1977.

Keefe, F. J., & Webb, J. T. Sex and the automated interview: Interviewer and interviewee sex difference effects. Paper presented at the Southeastern Psychological Association, March 1974, Hollywood Beach, Florida.

Kessler, J. W. *Psychopathology of childhood.* Englewood Cliffs, N.J.: Prentice-Hall, 1966.

Keutzer, C. S. Behavior modification of smoking. The experimental investigation of diverse techniques. *Behavior Research & Therapy,* 1968, *6,* 137–158.

Kohlenberg, R., & Phillips, T. Reinforcement and rate of litter depositing. *Journal of Applied Behavior Analysis,* 1973, *6,* 391–396.

Kohlenberg, R., Phillips, T., & Proctor, W. A behavioral analysis of peaking in residential electrical-energy consumers. *Journal of Applied Behavior Analysis,* 1976, *9,* 13–18.

Kopel, S. A. The relative efficacy of a self-control rapid smoking strategy for the maintenance of smoking reduction. Paper presented at the Ninth Annual Meeting of the Association for Advancement of Behavior Therapy, San Francisco, December 1975.

Kopel, S. A., & Arkowitz, H. Role-playing as a source of self-observation and behavior change. *Journal of Personality and Social Psychology,* 1974, *29,* 677–686.

Kopel, S. A., & Arkowitz, H. The role of attribution and self-perception in behavior change: Implications for behavior therapy. *Genetic Psychology Monographs,* 1975, *92,* 175–212.

Lacey, J. I. Psychophysiological approaches to the evaluation of psychotherapy process and outcome. In E. H. Rubinstein & M. B. Parloff (Eds.), *Research in psychotherapy.* Vol. 1. Washington, D.C.: American Psychological Association, 1959.

Lang, P. J. The mechanics of desensitization and the laboratory study of human fear. In C. M. Franks (Ed.), *Behavior therapy appraisal and status.* New York: McGraw-Hill, 1969.

Lang, P. J. The application of psychophysiological methods to the study of psychotherapy and behavior modification. In A. E. Bergin & S. L. Garfield (Eds.), *Handbook of psychotherapy and behavior change.* New York: John Wiley, 1971.

Lang, P. J., Melamed, B. G., & Hart, J. A psychophysiological analysis of fear modification using an automated desensitization procedure. *Journal of Abnormal Psychology,* 1970, *72,* 220–234.

Lazarus, A. A. *Behavior therapy and beyond.* New York: McGraw-Hill, 1971.

Lazarus, A. A., & Abramovitz, A. The use of "emotive imagery" in the treatment of children's phobias. *Journal of Mental Science*, 1962, *108*, 191–195.

Lazarus, A. A., & Davison, G. C. Clinical innovation in research and practice. In A. E. Bergin & S. L. Garfield (Eds.), *Handbook of psychotherapy and behavior change*. New York: John Wiley, 1971.

Leaf, W. B., & Gaarder, K. R. A simplified electromyographic feedback apparatus for relaxation training. *Journal of Behavior Therapy and Experimental Psychiatry*, 1971, *2*, 39–44.

Leitenberg, H. The use of single-case methodology in psychotherapy research. *Journal of Abnormal Psychology*, 1973, *82*, 87–101.

LeLaurin, K., & Risley, T. R. The organization of day care environments: "Zone" vs. "man-to-man" staff assignments. *Journal of Applied Behavior Analysis*, 1972, *5*, 225–232.

Lewinsohn, P. M. The behavioral study and treatment of depression. In M. Hersen, R. M. Eisler, & P. M. Miller (Eds.), *Progress in behavior modification*. Vol. 1. New York: Academic Press, 1975.

Lewinsohn, P., Alper, T., Johansson, S., Libet, J., Rosenberry, C., Shaffer, M., Sterin, C., Steward, R., & Weinstein, M. Manual of instruction for the behavior rating used for the observation of interpersonal behavior. Unpublished manuscript, University of Oregon, 1968.

Lewittes, D. J., & Israel, A. C. Responsibility contracting for the maintenance of reduced smoking: A technique innovation. *Behavior Therapy*, 1975, *6*, 696–698.

Liberman, R. P., & Bryan, E. A behavioral approach to community psychiatry. *Archives of General Psychiatry*, in press.

Liberman, R. P., & Davis, J. Drugs and behavior analysis. In M. Hersen, R. M. Eisler, & P. M. Miller (Eds.), *Progress in behavior modification*. Vol. 1. New York: Academic Press, 1975.

Liberman, R. P., Davis, J., Moon, W., & Moore, J. Research design for analyzing drug-environment behavior interactions. *Journal of Nervous and Mental Disease*, 1973, *156*, 432–439.

Liberman, R. P., DeRisi, W. J., King, L. W., Eckman, T. H., & Wood, D. D. Behavioral measurement in a community mental health center. In P. O. Davidson, F. W. Clark, & L. A. Hamerlynck (Eds.), *Evaluation of behavioral programs in community, residential and school settings*. Proceedings of the *Fifth Banff Conference on Behavior Modification.* Champaign, Ill.: Research Press, 1974.

Liberman, R. P., King, L. W., & DeRisi, W. J. Behavior analysis and therapy in community mental health. In H. Leitenberg (Ed.), *Handbook of behavior modification and behavior therapy.* Englewood Cliffs, N.J.: Prentice-Hall, 1976.

Libet, J., & Lewinsohn, P. M. The concept of social skill, with special references to the behavior of depressed persons. *Journal of Consulting and Clinical Psychology,* 1973, *40,* 304–312.

Lieblum, S. R., & Kopel, S. A. Screening and prognosis in sex therapy: To treat or not to treat? *Behavior Therapy,* 1977, *8,* 480–486.

Lindsley, O. R. A reliable wrist counter for recording behavior rates. *Journal of Applied Behavior Analysis,* 1968, *1,* 77–78.

Lipinski, D., & Nelson, R. Problems in the use of naturalistic observation as a means of behavioral assessment. *Behavior Therapy,* 1974, *5,* 341–351.

Lobitz, C. W., & LoPiccolo, J. New methods in the behavioral treatment of sexual dysfunction. *Journal of Behavior Therapy and Experimental Psychiatry,* 1972, *3,* 265–271.

Locke, H. J., & Wallace, K. M. Short marital-adjustment and prediction tests: Their reliability and validity. *Journal of Marriage and Family Living,* 1959, *21,* 251–255.

LoPiccolo, J., & Steger, J. C. The sexual interaction inventory: A new instrument for assessment of sexual dysfunction. *Archives of Sexual Behavior,* 1974, *3,* 585–595.

Lorr, M., Klett, C. J., & McNair, D. M. *Syndromes of psychosis.* New York: Pergamon, 1963.

Lubetkin, B. The use of a planetarium in the desensitization of a case of bronto- and astra-phobia. *Behavior Therapy,* 1975, *6,* 276–277.

Lubin, B. Adjective checklists for the measurement of depression. *Archives of General Psychiatry,* 1965, *12,* 57–62.

MacPhillamy, D., & Lewinsohn, P. M. The Pleasant Events Schedule. Unpublished manuscript. Eugene: University of Oregon, 1971.

Mahoney, M. J. *Cognition and behavior modification.* Cambridge: Ballinger, 1974.

Mahoney, K. Count on it: A simple self-monitoring device. *Behavior Therapy,* 1974, *5,* 701–703.

Mandler, G., & Sarason, J. B. A study of anxiety and learning. *Journal of Abnormal and Social Psychology,* 1952, *47,* 166–173.

Mannino, F. V., MacLennon, B. W., & Shore, M. F. (Eds.). *The practice of mental health consultation.* New York: Gardner Press, 1975.

Marks, I. M., & Gelder, M. G. Transvestism and fetishism: Clinical and psychological changes during faradic aversion. *British Journal of Psychiatry,* 1967, *113,* 711–729.

Martin, R. *Legal challenges to behavior modification.* Champaign, Ill.: Research Press, 1975.

Martin, W. T. *Writing psychological reports.* Springfield, Ill.: Charles C Thomas, 1972.

Martyn, M. M., & Sheehan, J. Onset of stuttering and recovery. *Behavior Research & Therapy,* 1968, *6,* 295–308.

Masters, W. H., & Johnson, V. E. *Human sexual response.* Boston: Little, Brown, 1966.

Masters, W. H., & Johnson, V. E. *Human sexual inadequacy.* Boston: Little, Brown, 1970.

McCullough, J. P., & Montgomery, C. L. A technique for measuring subjective arousal in therapy clients. *Behavior Therapy,* 1972, *3,* 627–628.

McFall, P. M. The effects of self-monitoring on normal smoking behavior. *Journal of Consulting and Clinical Psychology,* 1970, *35,* 135–142.

McFall, R. M., & Hammen, C. L. Motivation, structure, and self-monitoring: The role of nonspecific factors in smoking reduction. *Journal of Consulting and Clinical Psychology,* 1972, *37,* 80–86.

McHugh, G. *Sex knowledge inventory: Form Y: Vocabulary and anatomy.* Durham, N.C.: Family Life Publications, 1955.

McHugh, G. *Sex knowledge inventory: Form X (revised).* Durham, N.C.: Family Life Publications, 1967.

McHugh, G. *Marriage counselor's manual and teacher's handbook.* Durham, N.C.: Family Life Publications, 1968.

McPherson, F. M., & LeGassicke, J. A single-patient, self-controlled and self-recorded trial of WY 3498. *British Journal of Psychiatry,* 1965, *111,* 149–154.

Meehl, P. E. The cognitive activity of the clinician. *American Psychologist,* 1960, *15,* 19–27.

Meichenbaum, D. A cognitive-behavior modification assessment approach. In M. Hersen & A. Bellack (Eds.), *Behavioral assessment: A practical handbook.* New York: Pergamon Press, 1976.

Melamed, B. J., & Siegel, C. J. Self-directed in-vivo treatment of an obsessive-compulsive checking ritual. *Journal of Behavior Therapy and Experimental Psychiatry,* 1975, *6,* 31–36.

Meyers, H., Nathan, P., & Kopel, S. Effects of a token reinforcement system on journal reshelving. *Journal of Applied Behavior Analysis,* 1977, in press.

Miller, P. M. The use of behavioral contracting in the treatment of alcoholism: A case report. *Behavior Therapy,* 1972, *3,* 593–596.

Miller, W. H. *Systematic parent training.* Champaign, Ill.: Research Press, 1975.

Mischel, W. *Personality and assessment.* New York: John Wiley, 1968.

Mischel, W. On the future of personality measurement. *American Psychologist,* 1977, *32,* 246–254.

Morris, R. J., & Suckerman, K. R. Therapist warmth as a factor in automated systematic desensitization. *Journal of Consulting and Clinical Psychology,* 1974, *42,* 244–250.

Nathan, P. E., Goldman, M. S., Lisman, S. A., & Taylor, A. A. Alcohol and alcoholics: A behavioral approach. *Transactions of the New York Academy of Sciences,* 1972, *34,* 602–627.

Nathan, P., Schneller, P., & Lindsley, O. Direct measurement of communication during psychiatric admission interviews. *Behavior Research & Therapy,* 1964, *2,* 49–57.

Nelson, R. O., & Bowles, P. E. The best of two worlds—observations with norms. *Journal of School Psychology,* 1975, *13,* 3–9.

Nord, W. R. Beyond the teaching machine: The neglected area of operant conditioning in the theory and practice of management. *Organizational Behavior and Human Performance,* 1969, *4,* 375–401.

O'Leary, K. D. The assessment of psychopathology in children. In H. C. Quay & J. S. Werry (Eds.), *Psychopathological disorders of children.* New York: John Wiley, 1972.

Patterson, G. R. Behavioral intervention procedures in the classroom and in the home. In A. E. Berzin & S. L. Garfield (Eds.), *Handbook of psychotherapy and behavior change.* New York: John Wiley, 1971.

Patterson, G. R., & Gullion, M. E. *Living with children.* Champaign, Ill.: Research Press, 1971.

Patterson, G. R., & Hops, H. Coercion, a game for two: Intervention techniques for marital conflict. In R. E. Ulrich & P. Mountjoy (Eds.), *The experimental analysis of social behavior.* New York: Appleton-Century-Crofts, 1972.

Patterson, G. R., & Reid, J. B. Reciprocity and coercion: Two facets of social systems. In C. Neuringer & J. Michael (Eds.), *Behavior modification in clinical psychology.* New York: Appleton-Century-Crofts, 1970.

Patterson, G. R., Weiss, R. L., & Hops, H. Training of marital skills: Some problems and concepts. In H. Leitenberg (Ed.), *Handbook of behavior modification.* Englewood Cliffs, N.J.: Prentice-Hall, 1975.

Peterson, D. K. A functional approach to the study of person-person interactions. In D. Magnusson & N. Endler (Eds.), *Personality at the crossroads: Current issues in transactional psychology.* New York: John Wiley, 1977.

Pierce, C. H., & Risley, T. R. Improving job performance of Neighborhood Youth Corps aids in an urban recreation program. *Journal of Applied Behavior Analysis*, 1974, 7, 207–216.

Pion, R. J. *The sexual response profile.* Honolulu: Enabling Systems, 1975.

Powers, R. B., Osborne, J. G., and Anderson, E. G. Positive reinforcement of litter removal in the natural environment. *Journal of Applied Behavior Analysis.*, 1973, 6, 579–586.

Premack, D. Toward empirical behavior laws: I. Positive reinforcement. *Psychological Review*, 1959, 66, 219–233.

Proshansky, H. M., Ittelson, W. H., & Rivlin, L. G. (Eds.). *Environmental psychology.* New York: Holt, Rinehart and Winston, 1970.

Purcell, K., & Brady, K. Adaptation to the invasion of privacy: Monitoring behavior with a miniature radio transmitter. *Merrill-Palmer Quarterly,* 1966, 12, 242–254.

Rabavilas, A. D., & Boulougouris, J. C. Physiological accompaniments of ruminations, flooding, and thought stopping in obsessive patients. *Behavior Research & Therapy*, 1974, 12, 239–244.

Rachman, S., Hodgson, R., & Marks, I. M. The treatment of chronic obsessive-compulsive neurosis. *Behavior Research & Therapy*, 1971, 9, 237–238.

Rachman, S., & Hodgson, R. I. Synchrony and dysynchrony in fear and avoidance. *Behavior Research & Therapy*, 1974, 12, 311–318.

Rathus, S. A. An experimental investigation of assertive training in a group setting. *Journal of Behavior Therapy and Experimental Psychiatry*, 1972, 3, 81–86.

Rathus, S. A. A thirty-item schedule for assessing assertive behavior. *Behavior Therapy*, 1973, 4, 298–406.

Reid, J. B. Reliability assessment of observation data. A possible methodological problem. *Child Development*, 1970, *41*, 1143–1150.

Repucci, N. D., & Saunders, J. T. Social psychology of behavior modification: Problems of implementation in natural settings. *American Psychologist*, 1974, *29*, 649–660.

Risley, T. R., & Wolf, M. M. Strategies for analyzing behavior change over time. In J. Nesselroade & H. Reese (Eds.), *Life-span developmental psychology: Methodological issues.* New York: Academic Press, 1972.

Robinson, C. H., & Annon, J. S. *The heterosexual attitude scale—Male form.* Honolulu: Enabling Systems, 1975. (a)

Robinson, C. H., & Annon, J. S. *The heterosexual attitude scale—Female form.* Honolulu: Enabling Systems, 1975. (b)

Robinson, C. H., & Annon, J. S. *The heterosexual behavior inventory—Male form.* Honolulu: Enabling Systems, 1975. (c)

Robinson, C. H., & Annon, J. S. *The heterosexual behavior inventory—Female form.* Honolulu: Enabling Systems, 1975. (d)

Romanczyk, R. G., Tracey, D. A., Wilson, G. T., & Thorpe, G. L. Behavioral techniques in the treatment of obesity: A comparative analysis. *Behavior Research & Therapy*, 1973, *11*, 629–640.

Rose, M. Needed research on the mediation of labor disputes. *Personnel psychology*, 1952, *5*, 187–200.

Rosen, R. C., & Keefe, F. J. The measure of human penile tumescence. Unpublished manuscript, Rutgers Medical School, 1977.

Rosen, R. C., & Kopel, S. A. Penile Plethysmography and biofeedback in the treatment of a transvestite-exhibitionist. *Journal of Consulting and Clinical Psychology*, in press.

Rosenhan, D. L. On being sane in insane places. *Science*, 1973, *179*, 250–258.

Rosenthal, R. On the social psychology of the psychological experiment: The experimenter's hypothesis as unintended determinant of experimental results. *American Scientist*, 1963, *51*, 268–283.

Rosenthal, R. *Experimental effects in behavioral research.* New York: Appleton-Century-Crofts, 1966.

Ross, H., Campbell, D. T., & Glass, G. V. Determining the social effects of a legal reform: The British "Breathalyzer" crackdown of 1967. *American Behavioral Scientist*, 1970, *13*, 493–509.

Rugh, J. D., & Schwitzgebel, R. L. Biofeedback apparatus: List of suppliers. *Behavior Therapy*, 1975, 6, 238–240. (a)
Rugh, J. D., & Schwitzgebel, R. L. Biofeedback apparatus: List of suppliers. *Behavior Therapy*, 1975, 6, 423. (b)
Sanford, N. Personality: Its place in psychology. In I. S. Koch (Ed.), *Psychology: A study of a science*. Vol. 5. New York: McGraw-Hill, 1963.
Sarason, I. G. Interrelationship among individual difference variables, behavior in psychotherapy, and verbal conditioning. *Journal of Abnormal and Social Psychology*, 1958, 56, 339–344.
Schaefer, H. H., & Martin, P. L. *Behavioral therapy*. New York: McGraw-Hill, 1969.
Schaefer, H. H., Sobell, M. B., & Mills, K. C. Some sobering data on the use of self-confrontation with alcoholics. *Behavior Therapy*, 1971, 2, 28–39.
Schifferes, J. J. *What's your caloric number?* New York: Macmillan, 1966.
Schmidt, H. O., & Fonda, C. P. The reliability of psychiatric diagnosis: A new book. *Journal of Abnormal Social Psychology*, 1956, 52, 262–267.
Schnelle, J. F., Kirchner, R. E., McNees, M. P., & Lawler, J. M. Social evaluation research: The evaluation of two police patrolling strategies. *Journal of Applied Behavior Analysis*, 1975, 8, 353–366.
Schroeder, S. R. Automated transduction of sheltered workshop behaviors. *Journal of Applied Behavior Analysis*, 1972, 5, 523–525.
Schuessler, K. F. The deterrent influence of the death penalty. In W. J. Chambliss (Ed.), *Crime and the legal process*. New York: McGraw-Hill, 1969.
Schwartz, G. E. Biofeedback as therapy: Some theoretical and practical issues. *American Psychologist*, 1973, 28, 666–673.
Shapiro, D., & Surwit, R. S. Learned control of physiological function and disease. In H. Leitenberg (Ed.), *Handbook of behavior modification and behavior therapy*. Englewood Cliffs, N.J.: Prentice-Hall, 1976.
Shorkey, C. T. Behavior therapy training in social work education. *Journal of Behavior Therapy and Experimental Psychiatry*, 1973, 4, 195–196.
Sidman, M. *Tactics of scientific research*. New York: Basic Books, 1960.

Simmons, J. E. *Psychiatric examination of children.* Philadelphia: Lea & Febinger, 1969.
Skinner, B. F. *Beyond freedom and dignity.* New York: Knopf, 1971.
Sommer, R. The social psychology of the cell environment. In *New environments for the incarcerated.* Washington, D. C.: U.S. Department of Justice, Law Enforcement Administration, National Institute of Law Enforcement and Criminal Justice, 1972.
Stebbins, R. A. *The disorderly classroom: Its physical and temporal conditions.* St. John's, Newfoundland: Committee on Publications, Faculty of Education, Memorial University, 1974.
Stieber, J. W. *Ten years of the Minnesota Labor Relations Act.* Minneapolis: University of Minnesota, Industrial Relations Center, 1949.
Stoler, P. Exploring the frontiers of the mind. *Time,* 1974, *103* (2), 50–59.
Stuart, R. B. Operant-interpersonal treatment for marital discord. *Journal of Consulting and Clinical Psychology,* 1969, *3,* 675–682.
Stuart, R. B. Behavioral remedies for marital ills: A guide to the use of operant-interpersonal techniques. In T. Thompson & W. Docken (Eds.), *International symposium on behavior modification.* New York: Appleton-Century-Crofts, 1975.
Stuart, R. B., & Davis, B. *Slim chance in a fat world.* Champaign, Ill.: Research Press, 1972.
Stuart, R. B., & Stuart, F. *Marital pre-counseling inventory.* Champaign, Ill.: Research Press, 1972.
Suinn, R. M. The STABS, a measure of test anxiety for behavior therapy: Normative data. *Behavior Research & Therapy,* 1969, *7,* 335–340.
Sullivan, H. S. *The psychiatric interview.* New York: Norton, 1954.
Szasz, T. S. *The myth of mental illness.* New York: Hoeber-Harper, 1961.
Tate, B. An automated system for reinforcing and recording retardate work behaviors. *Journal of Applied Behavior Analysis,* 1968, *1,* 347–348.
Tasto, D. P., & Chesney, M. Muscle relaxation treatment for primary dysmenorrhea. *Behavior Therapy,* 1974, *5,* 668–672.
Taylor, J. A. A personality scale of manifest anxiety. *Journal of Abnormal and Social Psychology,* 1953, *48,* 285–290.

Tharp, R. G., & Wetzel, R. J. *Behavior modification in the natural environment.* New York: Academic Press, 1969.

Thibaut, J. W., & Kelley, H. H. *The social psychology of groups.* New York: John Wiley, 1959.

Thomas, E. J. Bias and therapist influence in behavioral assessment. *Journal of Behavior Therapy and Experimental Psychiatry,* 1973, *4,* 107–111.

Thomas, E. J., Carter, R. D., & Gambrill, E. D. Some possibilities of behavioral modification with marital problems using SAM (Signal system for the assessment and modification of behavior). In R. D. Rubin, H. Fensterheim, A. A. Lazarus, & C. M. Franks (Eds.), *Advances in behavior therapy.* New York: Academic Press, 1971.

Thoreson, C. E., & Mahoney, M. J. *Behavioral self-control.* New York: Holt, Rinehart and Winston, 1974.

Turner, M. The lessons of Norman House. *Annals of the American Academy of Political and Social Science,* 1969, *381,* 39–46.

Twardosz, S., Cataldo, M. F., & Risley, T. R. An open environment design for infant and toddler care. *Journal of Applied Behavior Analysis,* 1974, *7,* 529–546.

Ullmann, L. P., & Krasner, L. Introduction: What is behavior modification? In L. P. Ullmann & L. Krasner (Eds.), *Case studies in behavior modification.* New York: Holt, Rinehart and Winston 1965.

Ullmann, L. P., & Krasner, L. *A psychological approach to abnormal behavior.* Englewood Cliffs, N. J.: Prentice-Hall, 1969.

Verhave, T. The pigeon as a quality-control inspector. In R. Ulrich, T. Strachnik, & J. Mabry (Eds.), *Control of human behavior.* Vol. 1. Glenview, Ill.: Scott Foresman, 1966.

Vincent, J. P. The relationships of sex, degree of intimacy, and degree of mental distress to problem-solving behavior and exchange of social reinforcement. Unpublished doctoral dissertation, University of Oregon, 1972.

Vincent, J. P., Weiss, R. L., & Birchler, G. R. A behavioral analysis of problem solving in distressed and nondistressed married and stranger dyads. *Behavior Therapy,* 1975, *6,* 475–487.

Wagner, M. K. Reinforcement of the expression of anger through role playing. *Behavior Research & Therapy,* 1968, *6,* 91–96.

Walker, H. M. *Walker Problem Behavior Identification Checklist.* Los Angeles: Western Psychological Services, 1970.

Walker, H. M., & Hops, H. Use of normative peer data as a standard for evaluating classroom treatment effects. *Journal of Applied Behavior Analysis,* 1976, *9,* 159–168.

Walker, H. M., Hops, H., & Johnson, S. M. Generalization and maintenance of treatment effects. *Behavior Therapy*, 1975, 6, 188–200.

Watson, D. L., & Friend, R. Measurement of social-evaluative anxiety. *Journal of Consulting and Clinical Psychology*, 1969, 33, 448–457.

Watson, D. L., & Tharp, R. G. *Self-directed behavior: Self-modification for personal adjustment*. Belmont, Cal.: Brooks/Cole, 1972.

Watson, J. P., Gaind, R., & Marks, I. M. Physiological habituation to continuous phobic stimulation. *Behavior Research & Therapy*, 1972, 10, 269–278.

Watson, J. P., Mullett, G. E., & Pillay, G. E. The effect of prolonged exposure to phobic situations upon agoraphobic patients treated in groups. *Behavior Research & Therapy*, 1973, 11, 531–546.

Weed, L. L. Medical records that guide and teach. *New England Journal of Medicine*, 1968, 278, 593–600.

Weiss, R. L., Hops, H., & Patterson, G. R. A framework for conceptualizing marital conflict, a technology for altering it, some data for evaluating it. In L. A. Hamerlynch, L. C. Handy, & E. Mash (Eds.), *Behavior change: Methodology, concepts and practice*. Champaign, Ill.: Research Press, 1973.

Weiss, R. L., & Margolin, G. Marital conflict and accord. In A. R. Ciminero, K. S. Calhoun, & H. E. Adams (Eds.), *Handbook for behavioral assessment*. New York: John Wiley, 1975.

Wexler, D. B. Token and taboo: Behavior modification, token economies and the law. *Behaviorism: A Forum for Critical Discussion*, 1973, 1, 1–24.

Wiggins, J. S. *Personality and prediction: Principles of personality assessment*. Reading, Mass.: Addison Wesley, 1973.

Willems, E. P. Behavioral technology and behavioral ecology. *Journal of Applied Behavior Analysis*, 1974, 7, 151–165.

Winnett, R. A. Behavior modification and social change. *Professional Psychology*, 1974, 5, 147–153.

Wisocki, P. A. Treatment of obsessive-compulsive behavior by covert sensitization: A case report. *Journal of Behavior Therapy and Experimental Psychiatry*, 1970, 1, 233–240.

Wollersheim, J. P. Effectiveness of group therapy based upon learning principles in the treatment of overweight women. *Journal of Abnormal Psychology*, 1970, 76, 462–474.

Wolpe, J. *The practice of behavior therapy*. New York: Pergamon, 1969.

Wolpe, J., & Flood, J. The effect of relaxation on the galvanic skin response to repeated phobic stimuli in ascending order. *Journal of Behavior Therapy and Experimental Psychiatry,* 1970, *1,* 195–200.

Wolpe, J., & Lang, P. J. A fear survey schedule for use in behavior therapy. *Behavior Research & Therapy,* 1964, *2,* 27–30.

Wolpe, J., & Lazarus, A. A. *Behavior therapy techniques.* New York: Pergamon Press, 1966.

Yates, A. S., & Poole, A. D. Behavioral analysis in a case of excessive frequency of micturition. *Behavior Therapy,* 1972, *3,* 449–453.

Yonovitz, A., & Kumar, A. An economical easily recordable galvanic response apparatus. *Behavior Therapy,* 1972, *3,* 629–630.

Zifferblatt, S. M., & Hendricks, C. G. Applied behavioral analysis of societal problems: Population change, a case in point. *American Psychologist,* 1974, *29,* 750–762.

Zuckerman, M. The development of an affective adjective checklist for the measurement of anxiety. *Journal of Consulting Psychology,* 1960, *24,* 457–462.

Zung, W. W. K. A self-rating depression scale. *Archives of General Psychiatry,* 1965, *12,* 63–70.

Index

A-B-A-B withdrawal design, 28–30
Adult outpatients, behavioral assessment of
   automatic recording devices for, 82–83
   behavioral antecedents in, 68–69
   checklists, inventories, and surveys for, 70–71
   home observation in, 78–79
   the interview in, 64–68
   in vivo tests for, 80–81
   laboratory observation in, 83–85
   naturalistic observation in, 78–83
   psychophysiological analysis in, 87–90
   questionnaires for, 68–74
   rating scales for, 69–70
   response characteristics in, 69–71
   role playing in, 83–84
   self-observation in, 74–78
   simulations in, 84–85
Antecedents in functional analysis, 21–22
Assessment of ongoing therapy
   with children, 62–63
   in the framework of behavioral assessment, 26–28
   with institutionalized patients, 141–142
   in marital discord, 105–106
   in sexual dysfunction, 121–123

Behavioral-analytic strategy, 6
Behavioral assessment
   of adult outpatients. *See* Adult outpatients
   of children. *See* Children
   of community problems. *See* Community problems
   in educational systems. *See* Educational systems
   ethics of, 169
   evaluation of therapy in, 28–33
   functional analysis in, 20, 21–23
   in industrial systems. *See* Industrial systems
   of institutionalized patients. *See* Institutionalized patients
   the interview in, 9–10
   legal issues concerning, 169–170
   of living environments. *See* Living environments
   of marital discord. *See* Marital discord
   matching treatment to patient in, 23–26
   measurement of behavior in, 20–21
   in mental health fields, 168–169
   methods of, 9–12
   observation, in, 10–11
   and ongoing therapy, 26–28
   problem identification, 17–19
   psychological testing in, 11–12
   refinement of procedures of, 169
   research directions for, 170–171
   service directions for, 168–170
   of sexual dysfunction. *See* Sexual dysfunction
   of social and legal reforms, 159–160
   strengths of, 13–14
   of systems and society. *See* Systems and society
   teaching, 164–167
   technological advances in, 171
   theoretical foundation of, 4–7
   uses of, 8
   weaknesses of, 14–15
   See also Traditional assessment
"Black-box" model

Children, behavioral assessment of
  checklists and questionnaires for, 48–49
  classroom observation in, 59–60
  evaluation of therapy in, 62–63
  home observation in, 56–59
  matching treatment to client in, 60–62
  measurement and functional analysis in, 52–60
  observation by mediators in, 52–56
  and ongoing therapy, 62–63
  parent interview in, 37–46
  problem identification in, 37–51
  problem selection in, 50–51
  school personnel interview in, 46–48
Cognitive mediational responses in functional analysis, 22–23
Community problems, behavioral assessment of, 156–159
  conservation of energy resources, 157–159
  littering, 156–157
Consequences in functional analysis, 22
Consequences in S-O-R-K-C, 6
Contingency relationships in S-O-R-K-C, 6

Daily Report Form, 121–123
Diagnostic and Statistical Manual of Mental Disorders (DSM-II), 7–8
Diagnostic classification, 7
Direct sampling, principle of, 6

Educational systems, behavioral assessment in, 150–153
  misuse of educational material, 151, 152
  racial integration, 151, 153
  school discipline, 150
Evaluation of therapy
  for children, 62–63
  in framework of behavioral assessment, 28–33
  for institutionalized patients, 142–143
  in marital discord, 106
  in sexual dysfunction, 123–125

Family Questionnaire, 173–175
Fear Survey Schedule (FSS), 69
Functional analysis. *See* Measurement and functional analysis

Indicant phallacy, 89–90
Industrial systems, behavioral assessment in, 148–150
  job performance, 149–150
  tardiness, 148
  unemployment, 148
Initial caretaker interview, 39
Institutionalized patients, behavioral assessment of
  behavioral goal setting in, 133–134
  behavioral treatment program in, 135–142
  clinician's resources in, 139–140
  direct measurement of a permanent product in, 131
  drug treatment programs in, 134–135
  evaluation of therapy in, 142–143
  event recording in, 131–132
  institutional resources in, 140–141
  interviews in, 126–127
  legal issues in treatment in, 136–139
  matching treatment to client in, 133–142
  measurement and functional analysis, 129–133
  methods of observation in, 130–133
  and ongoing therapy, 141–142
  problem identification in, 126–129
  patient's resources in, 139
  preliminary observations in, 127–128
  problem listing in, 128–129
  questionnaires for, 127–128
  reviewing patient records in, 127
  target behavior in, 135–136
  time sampling in, 132–133
  Wyatt v. *Stickney* and, 136–139
Interviewing
  adult outpatients, 64–68
  institutionalized patients, 126–127
  marital discord clients, 91–93
  parents, 37–46
  school personnel, 46–48
  sexual dysfunction clients, 109–111

Living environments, behavioral assessment of
  activity measures, 153–154
  interaction measures, 154–155
  stimulation measures, 155–156
Locke-Wallace Marital Adjustment Scale (MAS), 49, 94–95, 115
Louisville Behavior Checklist (LBCL), 48–49

Marital Adjustment Scale. *See* Locke-Wallace Marital Adjustment Scale
Marital discord, behavioral assessment of
  affectionate exchanges in, 96–97
  behavioral model for, 96–98

behavior change attempts in, 97–98
behavior description in, 99–100
conflict-resolution tasks in, 99–101
evaluation of therapy in, 106
exchanges of positives and negatives in, 100
general communication skills in, 100
the initial interview in, 91–93
laboratory observations in, 99–101
matching treatment to client in, 101–104
measurement and functional analysis in, 96–101
naturalistic observations in, 101
and ongoing therapy, 105–106
problem analysis and resolution in, 100
problem identification in, 91–96
problem-solving behavior in, 97
questionnaires for, 93–96
self-recordings for, 98–99
Marital Pre-Counseling Inventory, 94–95
Marked item technique, 157
Matching treatment to client
 with children, 60–62
 in framework of behavioral assessment, 23–26
 with institutionalized patients, 133–141
 with marital discord clients, 101–104
 with sexual dysfunction clients, 118–121
Measurement and functional analysis
 with children, 52–60
 in framework of behavioral assessment, 19–23
 with institutionalized patients, 129–133
 with marital discord clients, 96–101
 with sexual dysfunction clients, 117–118
 in S-O-R-K-C, 6
Multiple baseline design, 31–32

Objective tests, 11–12
Observation
 of adult outpatients, 74–87
 behavioral and traditional, compared, 10–11
 of children, 52–60
 of institutionalized patients, 130–133
 laboratory, 21, 83–85, 99–101
 of marital discord clients, 99–101
 in measurement and functional analysis, 20–21
 naturalistic, 21, 78–83, 101

self-, 20–21, 74–78
Ongoing therapy. *See* Assessment of ongoing therapy
Operational definitions, principle of, 7
Organismic variables, 5
Output display, 87

Planned Activity Check (PLA-CHECK), 153–154
Pleased Events Schedule, 72
PLISSIT model, 120
Problem identification
 with adult outpatients, 64–74
 with children, 37–51
 determining, 17–19
 with institutionalized patients, 126–129
 with marital discord clients, 91–96
 with sexual dysfunction clients, 107–116
Problem list, 128–129
Problem-Oriented Medical Record, 128
Projective hypothesis and test, 11
Psychodynamic theory, 3–4
Psychological testing, 11–12
Psychophysiological analysis of adult outpatients, 87–90

Questionnaires
 for adult outpatients, 68–74
 family history, 173–175
 for institutionalized patients, 127–128
 for marital discord clients, 93–96
 multimodal life history, 72, 176–184
 for parents, 48–49
 rating scales for, 69–70
 sexual, 114–116
 for teachers, 49

Rating scales, 69–70
Reinforcement schedules, 6
Reinforcement Survey Schedule (RSS), 71–72
Response characteristics, 17–18
Responses in S-O-R-K-C, 5–6

Sexual dysfunction, behavioral assessment of
 Daily Record Form for, 121–123
 diagnostic classification definitions for, 108
 evaluation of therapy in, 123–125
 the initial interview in, 109–111
 integrating assessment information in, 119–120

Sexual dysfunction (*continued*)
   measurement and functional analysis in, 117–118
   and ongoing therapy, 121–123
   PLISSIT model for, 120
   problem identification in, 107–116
   questionnaires for, 114–116
   sexual learning history in, 111–114
   S-O-R-K-C in, 118–120
Sexual Interaction Inventory (SSI), 114, 115, 116
Sharing conference topic outline, 25–26
Signal processing unit, 87
Social learning theory, 4–7
Society. *See* Systems and society
S-O-R-K-C model
   described, 5–6, 8
   in sexual dysfunction, 118–120
Static analysis, 20, 54–56
Stimulus events in S-O-R-K-C, 5
Subjective discomfort scale (SUDS), 69–70
Systems and society, behavioral assessment of
   broad applications, 147–160
   ethical issues, 163–164
   practical issues, 161–163
   problems of application, 161–163
   theoretical issues, 160–161

Target behavior, 135–136
"Three-term contingency" data collection, 52–53
Traditional assessment
   methods of, 9–12
   strengths of, 12–13
   theoretical foundation of, 3–4
   uses of, 7–8
   weaknesses of, 13
Trait theory, 3–4
Transducer, 87

Walker Problem Behavior Identification Checklist (WPBIC), 49
Weiss-Patterson model of marital discord, 96, 97–98
Wyatt v. Stickney, 136–139